DR. BOT

Dr. Bot

Why Doctors Can Fail Us— and How AI Could Save Lives

Charlotte Blease

YALE UNIVERSITY PRESS
NEW HAVEN AND LONDON

Copyright © 2025 Charlotte Blease

All rights reserved. This book may not be reproduced in whole or in part, in any form (beyond that copying permitted by Sections 107 and 108 of the U.S. Copyright Law and except by reviewers for the public press) without written permission from the publishers.

All reasonable efforts have been made to provide accurate sources for all images that appear in this book. Any discrepancies or omissions will be rectified in future editions.

For information about this and other Yale University Press publications, please contact:
U.S. Office: sales.press@yale.edu   yalebooks.com
Europe Office: sales@yaleup.co.uk   yalebooks.co.uk

Set in Minion Pro by IDSUK (DataConnection) Ltd
Printed and bound in the UK using 100% renewable electricity at CPI Group (UK) Ltd

Library of Congress Control Number: 2025935742
A catalogue record for this book is available from the British Library.
Authorized Representative in the EU: Easy Access System Europe, Mustamäe tee 50, 10621 Tallinn, Estonia, gpsr.requests@easproject.com

ISBN 978-0-300-24714-5

10 9 8 7 6 5 4 3 2 1

To Rosemary and Catherine,
and
in memory of Victor and Henry,
with love

Contents

Introduction: The Ailing Appointment 1

1   Prognosis and Treatment Options 14

Part I   Accessing Care

2   Patient Pilgrims 39

3   Web-side Visits 57

Part II   Disclosing Our Symptoms

4   Doctor Deference 77

5   Pouring Our Hearts Out to Machines 99

Part III   Diagnosis and Treatment

6   Crafting Clinicians 123

7   The Dark Art of Medicine 143

8   Building Digital Doctors 165

9   The Shotgun Marriage of Man and Machine 188

CONTENTS

Part IV Empathy

10	Doctors Getting Deep	215
11	Humanizing Healthcare without Doctors	237
	Conclusion: Leaving the Appointment	259
	Notes	*277*
	Acknowledgments	*316*
	Index	*319*

Introduction

The Ailing Appointment

Every few days a tragedy on the scale of 9/11 strikes, yet hardly anyone notices. On a daily basis, the equivalent of four Airbuses—each carrying 170 people—fall out of the sky killing all the passengers onboard.

This death toll seems unthinkable. Although these figures don't make the 24-hour news cycle, the inconceivable is happening on American soil—not at the hands of terrorists or the aviation industry but at the hands of those to whom we entrust our health. Medical error is one of the leading causes of death in the United States, responsible for the fatalities of over a quarter of a million Americans annually.[1]

Deepening the incision into patient care, the findings are grave. One type of medical error is misdiagnosis. Every year in the US diagnostic error affects at least 1 in 20 patients or 12 million adults,[2] causing permanent disability or death to 795,000 people.[3] In the UK diagnostic error is gauged to occur in around 5 percent of primary care consultations, placing "several million patients" at risk annually.[4]

Some experts believe these diagnostic error rates are significantly underestimated.[5] Autopsy studies show that major mistakes in diagnosing the cause of death happen in 10 to 25 percent of cases.[6] Globally, most people will face a diagnostic error at least once in their lifetime, with rates likely far higher in low- and middle-income countries.[7] In the European Union, 22 of 30 million patients with rare diseases don't yet have a diagnosis, and 8 million wait an average of ten years to receive it.[8] A significant portion of primary care visits involve medically unexplained symptoms, underscoring frequent diagnostic challenges.

Errors don't just happen with rare diseases—they often involve common conditions like heart failure, pneumonia, and even urinary tract infections.[5] In 2019, Dr. David Newman-Toker, an expert in diagnostic error at Johns Hopkins University School of Medicine, led an investigation into the so-called "Big Three" illnesses which comprise vascular diseases, infections, and cancer.[9] His team examined more than 11,500 medical malpractice claims related to misdiagnoses at leading American health insurance companies. They found that errors in doctors' judgment accounted for 86 percent of all cases.

As we'll discover throughout this book, errors and misdiagnoses are not the only problems patients confront. Consider what happens before we even receive a diagnosis: how we access care. Even if we have medical insurance or live in a location with a health system free at the point of access, patients with higher burdens of disease face hidden hurdles to see a doctor. From transport problems to juggling the logistics of gig-economy jobs, from the concealed costs of clinic visits to the physical challenges of getting there, these concerns are seldom openly discussed.

When we do manage to put a foot in the clinic door, and overcome fears about doctor-bothering, communication in the clinic can undermine our care. Among strangers, the flow of conversation can vary considerably: sometimes chatter courses like a bounteous brook, other times it sputters like a faulty faucet. In medicine, the effects of communication breakdown can be significant and serious, even a matter of life and death.

Patients are not all treated equitably, and Hippocratic Oaths are sometimes hypocritical oaths. In this book, you'll discover how gaps in medical education and hidden biases in face-to-face consultations can leave some of us at a disadvantage in the clinic. A hidden hierarchy of symptoms and disease can also influence the seriousness with which doctors treat our problems. Added to this, doctors' empathy is in short supply. For good reasons, clinicians' bedside manner can be minimalist—more like Dr. Spock than the kindly, caring but fictional Dr. Quinn (*Medicine Woman*). Empathy is also unevenly distributed—some of us receive more of it than others.

Medicine's second casualty

Patients are the victims of medicine's hidden ailments, but doctors are its second victims. The idea that clinicians are casualties of healthcare sounds like a medical sob story: we expect our well-remunerated, well-educated, and well-to-do doctors to get it right. However, the harder doctors work to make us better, the sicker they can make themselves. Underneath the commanding professional garb physicians face challenging realities that patients seldom see.

Doctors are exhausted. In a recent Medscape survey, half of all US doctors say they are burned out, with 20 percent reporting they are depressed.[10] In the UK, nearly half of all doctors report struggling to "'provide sufficient patient care' on at least one occasion each week" and 42 percent feel unable to cope with their weekly workload.[11]

It's worth saying that these figures reflect improvements in rates of burnout following the return to "normality" after the COVID-19 pandemic. During that time, from April 2020 to mid-2021, the surge of stress was overwhelming as doctors witnessed the devastating toll of illness and death—not only among patients but also their colleagues. In the US, four in ten doctors were depressed or anxious.[12] Nearly seven in ten Canadian doctors suffered burnout, and one in five considered quitting or did so.[13] In the UK, physician calls to the British Medical Association's mental health helpline increased by 80 percent.[14]

Doctors have higher rates of depression and suicide than the general public.[15] In the US alone, it is estimated that between 300 and 400 doctors take their own lives every year—the equivalent of one medical school graduating class dying by suicide annually.[16] Despite having legal protections for their licensure, many hesitate to disclose mental health conditions due to fears of stigma: nearly 40 percent of US physicians say they are reluctant to seek medical help out of fear of losing their job.[17]

Amid all these mounting pressures, patient numbers are surging. The world's population is aging at an ever-faster rate.[18] The United Nations forecasts that by 2037 we will share the planet with a billion more people.[19] Meanwhile the World Health Organization estimates that between 2020 and 2050, the number of people aged 80 years or older will triple to 426 million.[18]

Longer life and more people carry consequences for doctors and our healthcare systems. Consider the US: by 2050, most people aged 50 or older will live with one or more chronic illnesses including cancer, coronary heart disease, diabetes, arthritis, and Alzheimer's disease.[20] In the UK it's estimated that one in three people born today will suffer from some form of dementia in later years.[21] Debilitating illnesses require long-term medical attention and care.

Yet doctors are officially a scarce resource: we are not producing enough of them to meet patients' needs. The World Health Organization predicts that the global health workforce will grow to 53.9 million by 2030. This falls far short of the projected need for 80 million health workers, leaving a staggering gap of 18 million in low- and middle-income countries.

The Association of American Medical Colleges forecasts a shortage of up to 139,000 physicians by 2033.[22] Europe is running out of doctors too.[23] In France, nearly 7 million people do not have a GP.[24] In the UK, half of trainee doctors, one in three GPs, and one in four specialty doctors have considered leaving medicine to pursue an alternative career path.[25]

INTRODUCTION

Health systems are crumbling under patient pressures. Not only the systems and environments in which doctors work but doctors' thinking within these environments counts too. Greater burdens on fewer doctors create the perfect petri dish for errors.

Studies show that burnout is linked to increased rates of error, harm, and patient death.[26] In an extensive study, conducted over two years, physician burnout was strongly, though understandably, associated with a reduction in professional effort.[27] Fatigue and sleep deprivation also undermine the accuracy of clinical judgments.[28] Doctors who are stressed are two to three times more likely to admit suboptimal practices, including making more mistakes.[29] Depression is associated with higher rates of medical error, including delays in care, documentation errors, and mistakes in drug prescriptions.[30] One study found a sixfold increase in medication errors among doctors with depressive disorders.[31]

With palpable shortages of doctors and increasing waiting times for patients, many of us are profoundly frustrated. Although it might seem unfair to point the finger at patients, occasionally the frail, injured, and sick can add to doctors' strain too. Globally, around half of doctors have experienced some kind of patient-driven ill-treatment in the line of duty.[32] In the US, patient-initiated bad behavior contributes to doctors' stress and job turnover.[33] Meanwhile, in the UK, a survey of more than half a million National Health Service England staff found nearly one in four had experienced harassment, bullying, or abuse from patients.[34] Since clinician burnout is associated with high rates of error, patients may inadvertently be sabotaging their own care, and potentially that of other people.

Under pressure to placate anxious patients, and out of litigation fears for negligence or error, over 90 percent of American doctors admit to practicing "defensive medicine."[35] This occurs when doctors over-test and over-treat patients. By ordering a battery of unnecessary tests, such as scans and X-rays, doctors not only generate financial wastage totaling tens of billions of dollars every year, but also risk causing serious harm.[36] An estimated 80 million CT scans are performed in the US every year.[37] For patients who have undergone multiple CT scans, the risk of radiation-induced cancer due to excessive screening can be as high as 12 percent.[38]

When all-too-human errors do arise, patients don't tend to sue.[39] Still, around a third of doctors in the US will face at least one malpractice suit in their career.[40] Primary care physicians face more legal challenges than any other specialists; most cases are the result of errors resulting from delayed and missed diagnoses.[41] In a clinical catch-22, legal woes compound clinicians' stress, increasing the likelihood of medical error.

Medicine diagnosed

Health systems are creaking, and patients bear the brunt of the strain. Medical doctors gulp down many bitter pills too. Improved working conditions—more clinicians, less stress, better facilities—could alleviate many of medicine's maladies. But the next home truth may be more surprising: such strategies will be very far from a cure-all.

Even in the Shangri-La of current healthcare systems, with an over-abundance of staff, time, and limitless resources, medicine would still fail far too many. Doctors would still be prone to errors, patients would still be vulnerable to unfair treatment, and systems would still be destined to disappoint.

Already readers may worry that this book defends a laissez-faire approach to medicine. Nothing could be further from the truth. The reader will find no justifications in the pages that follow for neglecting investment in superior healthcare. The better the working conditions, the better the outcomes for everyone. However, this would also miss a fundamental point: current health systems radically fail too many of us.

Consider physical access to care. Bricks-and-mortar clinics can shape the definition of who is considered a privileged patient.

INTRODUCTION

Hospitals and clinics are typically located in densely populated areas. Historical happenstance has left specialists clustered in certain regions of the world. Geography carries real consequences for access. Too often it is the most chronically ill and impoverished patients who remain the least well-served.

Even after we obtain access to doctors, human psychology exerts a ceiling effect on the quality of care we receive. Human brains were not optimally adapted for the booming buzzing world of modern medicine. Like the rest of us, physicians have Stone Age minds, and our evolutionary heritage blots its messy thumbprints all over our time-capped consultations. Whether as a clinician or a patient, we enter the consulting room equipped with a suite of inbuilt algorithms sculpted for life on the savannahs of Africa. Our ancient psychology is primarily influenced but not implacably determined by the four fs: feeding, fleeing, fighting, and reproduction. This means that within highly pressurized visits, patients are not the perfect witnesses to, or communicators about, their medical problems and doctors are not always the ideal extractors of them.

Before any doctors reading this book throw it against the wall: modern medicine is one of humankind's most impressive accomplishments. In some ways it is a victim of its own success: more of us are living longer and being treated for a greater number of chronic conditions. Moreover, the expertise of doctors, especially when working in dire conditions, deserves to be recognized and applauded—in this book you will find it celebrated.

As patients, we often overestimate our doctors' abilities, unfairly blaming them for failing to keep up to date, messing up when things go wrong, or failing to show us compassion. But Mother Nature was not a pushy parent whose primary goal was to mold *Homo sapiens* into top-notch doctors. Reminding us of our evolutionary heritage, I refer to human doctors as *Homo medicus*.

Despite doctors' dedication, diligence, and hard work, evolution ripped ancestral scars on human behavior. We tend to like people who are like us. We give prolonged, preferential attention to those who have more of the good stuff—whether skills, good looks, or fine things. We struggle to give sustained attention to the have-nots, the elderly, the disabled, and the down-on-their luck. Despite our instincts to think better of ourselves—both as doctors and as patients—this book reveals our Paleolithic un-PC tendencies.

We will discover the multiple hazardous ways that the human mind is mismatched with modern clinic environments. From the accessibility of care to mutual first impressions in the consulting room; from how, or even whether, we talk about our symptoms to our doctors' inability to keep up to date; and from their professional manners with colleagues to their capacity for empathy with patients, *Homo sapiens'* cognitive heritage weaves threads that fray clinical outcomes.

To genuinely improve healthcare access, reduce harm, and save lives, we must fundamentally overhaul its delivery. Undertaking this task first requires a no-holds-barred, resolutely honest approach about human limitations.

In short, if the function of medicine is accessible and reliable care, the key question is who or what can perform that goal best. This book argues that doctor-mediated care in traditional clinic settings is the incumbent in a system ripe for revolution. As the title suggests, I explore the ways in which technology could present various solutions to some of healthcare's most hushed-up problems. The book will open a taboo conversation: how and why our very *human-ness* currently undermines the delivery of medicine and how artificial intelligence (AI) could improve it.

The book is not glibly critical of doctors, nor is it blindly optimistic. Readers expecting a gushing love letter to technology will be disappointed. I am not sanguine about digital tools. Implementing more AI in healthcare requires careful consideration of many factors and new trade-offs. I also acknowledge the seasons of hype and dashed hope in the historic almanac of AI, and the myopia of Silicon Valley's mantra to "move fast and break things." However, I confront the clear

INTRODUCTION

trajectory of technological change. Take the adoption of OpenAI's ChatGPT, which is widely regarded as one of the fastest user growth rates in tech history. AI, I argue—*thoughtfully implemented*—could offer our best hope for patients.

Furthermore, the world of digital innovation, just like the medical profession, commits the same cardinal sin: it neglects a closer examination of human psychology. Back in 2024, Kevin Weil, chief product officer at OpenAI, stated, "We want to make it possible to interact with AI in all the ways you interact with another human being."[42] In this book, we'll question whether creating AI in the image of man is always a smart idea. Enthusiasm for rapid product placement may overshadow how these tools could be better designed for specific patient needs. We'll also probe AI's shortcomings in its application to medicine.

Shortly we will begin our dissection and analysis of the messy entrails of the patient–doctor appointment. But before we put medicine on the gurney, a word about my own credentials to perform this overdue operation.

Why this book?

There are numerous distinguished thinkers—including academic medical doctors, AI experts, and futurists—writing about the role of technology in healthcare. Many point to the need for technological change from the perspective of economics, efficiency, and workplace constraints. However, this is the first book to consider, in a deep way, the diverse range of physical and psychological pitfalls associated with traditional medical visits. To justify a radical overhaul in the delivery of medical care, we must keenly understand human limitations and assess whether current technologies can effectively address the challenges.

My fiercest critics may assume that only doctors are well placed to write a book about medicine—after all, they serve on its front lines. I will say more about this in the next chapter, but for now, let me stress—this would be a serious mistake. Doctors are not experts on all things medical any more than clerics are experts on all things religious, lawyers on all things legal, or porn stars on all things sex.

No matter how astute doctors' observations, how sensitive and introspective their writing, and how well they satisfy our itch for gossip, even the most compelling medical narratives must come with a "buyer beware." Doctors' vision of what it is they do, and how, is also compromised by conflicts of interest. Besides, for very good reasons (they are too busy with the day job), physicians tend to lack domain expertise when it comes to the subterranean, psychological story of what it is they do exactly.

My background is philosophical psychology, and I work as a health informaticist. The latter is a field that combines psychology, social sciences, ethics, artificial intelligence, and digital innovations to understand how medical information is processed and used in patient care. For nearly twenty years I have been thinking and writing about healthcare in a genuinely interdisciplinary way. My perspective is international in scope, as I have worked in diverse and varied university departments across the UK, Ireland, Europe, and the US, including at Harvard Medical School.

Although I count physicians among my closest colleagues, I have no implicit allegiance to the medical profession, its social mores and habits, or even its long-term preservation. This outsider perspective is crucial to appraising traditional medicine's value.

Nor am I hemmed in by the pockets of Big Tech. When it comes to the virtues and vices of technological innovations, I am free from conflicting loyalties or competing interests. I have no shares or investments in Silicon Valley. In short, I am not financially committed to any preordained conclusion that current technology can answer medicine's problems.

In summary: I have no dog in the fight—except for the interests of patients, far too many of whom are underdogs, marginalized, or an afterthought in traditional care.

INTRODUCTION

My perspective is also personal. I have two siblings who live with an incurable, rare form of muscular dystrophy, one of whom waited two decades to be diagnosed. In 2023, while writing this book, I lost my beloved partner—and veteran *Guardian* journalist—Henry McDonald to cancer and six months later, my father, Victor Blease, to dementia. I witnessed first-hand the sheer brilliance of doctors and the kindness of health professionals. But I also observed how things can go wrong with care. Like so many of you who are reading these words, I recognize that medicine is a profoundly personal matter. The "personal" refers not merely to doctors and their jobs, but patients, loved ones, and lives.

Socrates depicted himself as a "gadfly"—someone who challenges the status quo and opens debate but remains an outsider. In the same spirit, the aim of this book is to foster a thoughtful conversation about the past and future of medicine. As a philosopher I understand the need for measured discussion. Necessarily, the book is written with a splinter of ice in the heart. American philosopher Wilfrid Sellars argued, "the aim of philosophy is to understand how things in the broadest sense of the term hang together in the broadest sense of the term."[43] In the pages that follow, I endeavor to do the same.

A fair, nonpartisan yet patient-centric perspective is vitally important to present an accurate appraisal of the present and future state of medicine. Identifying what technology does well and where it currently misses the mark—the tasks it currently excels at, or even surpasses humans on, and vice versa—requires an impartial vantage point. I am afforded a luxury that few doctors enjoy—the opportunity to take a step back and offer fresh insight.

The structure of the book

The book is bold in scope, yet inevitably crucial topics lie beyond its reach. The ethical adoption of AI invites a suite of pressing national and global concerns, from patient privacy and the misuse of sensitive health data to resource allocation and environmental sustainability, and from regulating digital tools and data governance to legal accountability when things go wrong. The ability of healthcare systems to work together—where different computer programs can easily share and understand each other's data—is another fundamental issue. Readers seeking insights on these other subjects should turn to other works and volumes.

In the pages that follow, you'll meet real patients and doctors whose stories bring this book to life. While their personal stories are rooted in specific places, professions, and health systems, the themes are universal. By stepping back, we'll uncover a shared logic—a universal grammar—that underpins the experiences of patients and doctors. In a few instances, names and details have been changed to protect privacy.

Doctors are keenly attuned to the pressures they face and those on the health systems in which they work, yet some problems are—for good reasons—masked from our doctors' view. Therefore, in the next chapter, we'll begin by exploring those more concealed instances when the medical community has a history of ignoring, belittling, or even repressing signs of its own infirmity. Like many patients who experience "symptom denial" physicians often argue that some problems with the visit are overstated, overblown, or just an inevitable fact of life. I explain why patients should be cautious about venerating medicine's tight grip and professional monopoly over the delivery of their care.

Following this, the book maps out the familiar trajectory of the visit, focusing on primary care. Taking a scalpel, we'll puncture through the familiar appearance of the routine, 10- to 20-minute medical consultation.

The story is in four main parts: Accessing Care, Disclosing Our Symptoms, Diagnosis and Treatment, and Empathy. Each part opens by describing a range of problems associated with traditional doctor-mediated care and closes by addressing the current capabilities of technology to address these problems. Aside from patients and doctors, you will meet psychologists, cognitive scientists, anthropologists, sociologists, tech innovators, philosophers, and a variety of world-leading experts along the way. We conclude by looking at what happens when we leave the appointment behind us.

I now invite readers to roll up their sleeves, scrub their hands, and join me in a graphic, unflinching dissection of the heart of healthcare.

1

Prognosis and Treatment Options

"Once I said to a patient, 'I'm not God,' but I really didn't believe that."

These are the words of the late Dr. Edward Rosenbaum, author of the straight-talking memoir *A Taste of My Own Medicine*.[1] Rosenbaum's admission sounds rather arrogant. However, in medicine personal conviction and self-assurance are necessary to make swift decisions. Those charged with saving lives can't succumb to the "*ums*" and "*ahs*" of the academic researcher or dilettantish debater.

Although they must act decisively, doctors are also required to tread the fine line between humility and hubris. Both under- and over-confidence carry costs in the clinic. Worryingly, over-confidence seems to swell when doctors are confronted by especially complex cases.[2] One study found hospital doctors overestimated the quality of care they provided to 60 percent of their patients in the hours prior to the emergence of life-threatening complications.[3] Or take an investigation into patients who died in intensive care units which found that doctors who were "completely certain" of the diagnosis were wrong 40 percent of the time.[4] Another study reported, "increased experience was associated with decreased likelihood of requesting second opinions, curbside consultations [seeking informal advice from a colleague], and reference materials, regardless of diagnostic accuracy."[2]

Evolutionary psychologist Robert Trivers has devoted his career to investigating the origins of self-deception, and says over-confidence was adaptive in our ancestral environments, creating a host of social advantages, such as skills of persuasion and status.[5,6] While the lie cannot be too big, if a person harbors an inflated opinion of themselves, they are more effective at convincing others of its truth. The gap between the environments in which these capacities evolved and the world of modern medicine helps to explain why our old psychology exerts undesirable effects.

Dr. Rosenbaum couldn't help but perceive himself as a deity in a white coat, and making matters worse, as patients, we desire doctors who are confident and authoritative. But the snag, as Trivers and subsequent research has shown, is that doctors' conviction is an unreliable surrogate for clinical accuracy.[2,7] As one medical researcher summed it up, "physicians are 'walking ... in a fog of misplaced optimism' with regard to their confidence."[8]

As we'll discover, this over-confidence is also clear in the profession's failure to fully confront doctors' human limitations. It also permeates medicine's protective stance against external interference in what it deems its rightful domain.

Symptom denial

In 1759, French Enlightenment philosopher Voltaire published his masterpiece *Candide*. The novel follows the young and naïve titular character on his travels where he faces a series of misfortunes and calamities. No matter how dire or imperfect Candide found the world to be, his tutor Dr. Pangloss advised him that the imperfections were justified, since "All is for the best" in "the best of all possible worlds."[9] Voltaire used the novel to critique the hypocrisy and corruption of institutions, governments, and individuals.

The medical profession has undoubtedly perpetuated a Panglossian attitude about some of medicine's own ailments, including how they can be successfully treated. This isn't to say that doctors are unaware of the grave challenges contemporary healthcare faces. Nor is it to suggest that doctors fail to recognize the importance of innovation and change. Throughout the book, we'll meet many visionaries who are striving to make healthcare better. However, at an individual, historical, and institutional level, there is a strong track-record of medicine burying its head and stethoscope in the sand, with serious consequences for patient care.

Consider diagnostic error. For most of the profession's history, the problem has been compartmentalized and marginalized. As an organized field, the science of diagnostics is, remarkably, a millennial: younger than Uber, iPads, and *Keeping Up with the Kardashians*. Unlike cancer, heart disease, and mental illness, the field never made the healthcare hit parade, with scant funding earmarked to explore what really goes on.

For example, in the US, a major report by the Institute of Medicine entitled *To Err Is Human* (1999) all but ignored the problem of misdiagnosis. Astonishingly, in the 270-page book-length investigation it mentioned "diagnostic error" twice.[10] In the wake of the report, a study comparing doctors' and patients' views on medical errors found that physicians were much less likely than the public to believe that the quality of care is a problem (29 percent vs 68 percent) or that a national agency was needed to address medical errors (24 percent vs 60 percent).[11]

In 2009, patient safety experts Professors David Newman-Toker and Peter Pronovost of Johns Hopkins School of Medicine bluntly noted in a landmark article in the *Journal of the American Medical Association*, "Diagnostic errors often are unrecognized or unreported, and the science of measuring these errors (and their effects) is underdeveloped."[12] Echoing this complaint, in 2010, Professor Robert Wachter, a leading authority on patient safety, railed, "Diagnostic errors don't get any respect."[13]

The first organization devoted to promoting research, awareness, and education about diagnostic errors (the Society to Improve

Diagnosis in Medicine) was officially launched in 2011.[14] Just a few years later, the earliest major report on diagnostic error commissioned by the Institute of Medicine in the US was published. Entitled *Improving Diagnosis in Health Care*, it can be viewed as the first step towards inaugurating the field.[15] Today, a diverse group of international researchers actively tackles the issue of misdiagnosis, advancing the science of clinical judgment. However, it's also fair to say this remains a fringe field.

Some readers—especially clinicians—may worry that this analysis is too critical. To this I reply, doctors are suspect self-analysands. In this they are not alone. Social psychologist Timothy Wilson of the University of Virginia argues, in his appositely titled book, that we humans are *Strangers to Ourselves*.[16] Although we have the subjective feeling that we know ourselves better than anyone else, our subconscious was not adapted to afford scrupulous, laser-like powers of psychological analysis. This fact can be difficult to accept but a mountain of research bears it out.[5,6] Comparing ourselves with others, we tend to marginally overestimate our attractiveness, abilities, and popularity.[16] Wilson dubs these tendencies the "Lake Wobegon Effect," from Garrison Keillor's fictional town where everyone, including the children, is above average.

Medicine has colonized Lake Wobegon and doctors have a lakeside view. Physicians—just like the rest of us—are Pollyannaish when it comes to self-awareness. This is why their own accounts of what they do, and how well they do it, should be heavily caveated.

For example, lack of professional insight is evident in doctors' response to the very idea that diagnostic error is a problem to be solved. The medical community often depicts error rates as overstated, overblown, or just an inevitable fact of life. Repressing and denying knowledge of its own infirmity, every major investigation into the prevalence of misdiagnosis to date has prompted a flurry of physician fulmination in top medical journals.[17-19] Doctors even blame their peers when confronted with negative data: surgeons, for example, underestimate the mortality rate of their patients compared with that of their colleagues.[20] A survey published in *The New England Journal of Medicine* concluded that the momentum for reducing medical error comes from outside the institution of medicine and "is sustained primarily by a range of groups and by the media's interest in the problem—not by practicing physicians or the public."[21]

Symptom denial is to some extent perpetuated by medicine's own mythologies. Bookstore shelves are bowed with medical memoirs penned by scribblers-in-scrubs. Such first-hand testimonies, many of them page-turners, are invaluable reports from frontline medicine, the "warts and all" frankness adding to their credibility. We are reminded that, through relentless blood, sweat, and tears, doctors too are human. The narrative intonation is often that of the reluctant hero.

Patients, in contrast, are depicted a bit like comic-book damsels. After sounding the alarm, trusted physicians swoop to the scene, their white coats unfurling behind them like superhero capes. Triumphing over the graft and grind of the day job, he or she is, like Superman, "Just doin' my job, ma'am." The imagery is not mine. For example, in his 2017 biography and bestseller, *This Is Going to Hurt*, British writer and former physician Adam Kay refers to the "hero complex all doctors pretend they don't have: Batman with a bleeper."[22]

The Übermensch mentality is also captured in Henry Marsh's blisteringly honest memoir *Do No Harm*. Standing in a long checkout queue at a supermarket he muses, "'And what did you do today?' I felt like asking them, annoyed that an important neurosurgeon like myself should be kept waiting after such a triumphant day's work."[23] Similarly, a historian told me that, mingling with the other delegates at an academic conference, one physician introduced himself by saying, "I save lives for a living, what do you do?" The incident was described as even more startling because the uppity delegate was, in fact, attending a symposium on medical history.

As viewers, readers, and consumers of popular culture, we too collude in a sentimentalized, oversimplified, and idealized image of medicine—a myth that doctors ought to be omniscient on all things medical. Adam Kay reflects, "patients don't actually think of doctors as being human. It's why they're so quick to complain if we make a mistake or if we get cross ... it's the flip side of not wanting your doctor to be fallible."[22] Society expects doctors to be demigods.

Defying advice

Regrettably, like a defiant patient refusing treatment, the medical profession often resists input from outsiders—patients, other experts, or even groundbreaking innovations that have the potential to save lives. Understanding the roots of this resistance is as critical as it is complex. Here, a historical perspective helps.

In his book, *Bad Medicine: Doctors Doing Harm since Hippocrates*, British historian David Wootton argues that medicine has myopically overlooked key insights from thinkers and ideas located in adjacent fields.[24] From antiseptics to anesthesia, from the germ theory of disease to handwashing, from cell biology to clinical trials, and from penicillin to vaccinations, the medical community underestimated the significance and clinical value of advances that we, and it, now take for granted. Even basic tools like blood pressure monitors were once met with disdain and skepticism.

Wootton recognizes the complexities of medicine's journey and acknowledges that what constitutes "good evidence" has advanced with the passage of time. Nonetheless, even understood by the standards of its day, medicine has repeatedly failed to engage with key innovations. Consider anesthesia: doctors resisted it in part because they felt it undermined their surgical skills, which had been honed to operate quickly and efficiently. Confronted with a choice, and despite the literal pains and cries of patients, surgeons feared diminishing the value of their hard-earned expertise.

Strange as it sounds, and in defense of doctors, such conservatism isn't always a bad thing. There can be method in a modicum of maddening stubbornness. In his account of *The Structure of Scientific Revolutions*, twentieth-century historian Thomas Kuhn described an inbuilt defensiveness among scientists in upholding old ideas and practices. He observed that such responses protect against fads, hype, and bounding onto bandwagons.[25] Undoubtedly, in medicine too, changes to practice can be disruptive and there must be good reasons—a high bar—to radically overhaul traditional ways of working.

Another reason the profession resists change is the pressure of the job. As we'll find out in the book, doctors already battle relentless workloads. Reshaping their workflows is understandably a nightmarish proposition. And as we'll see later, many technological innovations are developed without taking this into consideration.

Yet solving serious problems in healthcare can and should serve as a powerful catalyst for change. For transformation to occur, doctors and medical organizations must recognize and respond to the pressing need for adaptation. When the motivation to evolve is absent, this often becomes yet another barrier to progress, something that leading voices in the profession acknowledge. For example, back in 2010, Richard Smith, former editor of the *BMJ* (*British Medical Journal*), inquired in his blog: "Computers take histories better than doctors—why don't they do it more?" His answer: "we don't change until we have to, and we haven't yet had to."[26]

Doctors' uptake of electronic communications has been slower than patients' willingness to use them. For example, in November 2015, nearly 25 years after the internet got going, *The Washington Post* inquired, "Want to reach your doc? Many Americans would use email or text—but can't": only 15 percent were able to use secure online email to communicate with their provider.[27] By 2021, an estimated 70 percent of US providers still riskily relied on facsimile (fax) machines to share patients' health information.[28]

Across the pond, by 2019 the UK's National Health Service (NHS) was still relying on its 8,000 fax machines for communication, and totting up £100 million a year on stamps and paper to mail letters to patients.[29] Despite announcements that the archaic fax machines would be phased out for NHS or patient communications, progress is glacial. In 2023, newspapers reported that around 600 fax machines were still in operation across the NHS.[30]

If the pace of change is painfully unhurried—often for understandable reasons—a further encumbrance to progress is medical education and what doctors learn. Physician and writer Danielle Ofri observes that the chief priority of medical education is to stuff biomedical facts into students as if it were force-feeding geese for foie gras.[31] To extend Ofri's vivid analogy, the compulsory force-feeding of biomedical facts at the expense of other fields leads to intellectual immobilism.

Educators and students already have a lot to cover. However, intensive education in psychology, cognitive science, artificial intelligence, and other fields that are relevant to contemporary medicine have largely been marginalized. Although some elite medical schools incorporate such courses—often taught by medical doctors rather than by domain experts—these subjects reside firmly on the fringes. Equally, some leading figures describe medicine as an "information processing" field. However, few pay close attention to unpacking exactly what this means. As a result, doctors' insights into the *hows*, *whys*, and *wherefores* of the psychology of what they do is limited. Again, this is not to stick the surgical knife into doctors' backs; rather it is to recognize an unintended consequence of doctors' education.

Pollyannaish optimism extends to thinking deeply about how—if at all—the profession might strengthen and improve. To invoke Ofri's dietary metaphor, when it comes to remedying medicine's ailments, some educational and training strategies are more akin to homely chicken soup: comforting but far from a cure. Throughout the book we will see that, from anti-bias training to courses in creative writing, left to treat its own maladies too many of medicine's antidotes are weak and inadequate, amounting to wishful thinking. Staggeringly, a recent review of interventions employed to tackle misdiagnosis reported patient outcomes were rarely measured.[32] The same is true for discussions about enhancing empathy in clinical care, where the effectiveness and appropriateness of various training interventions have scarcely been considered.

The perils of self-healing

If medicine has a history of overlooking outsider input, there are more ominous risks in relying on the health profession to treat its own ailments. Professional interests don't always align with those of the patients who rely on the services they provide.

Panning out to take a panoptic view of white-collar work is instructive. In their book *The Future of the Professions*, father-and-son authors and Oxford University academics Richard and Daniel Susskind discuss the concept of a "grand bargain" that exists between society and the professions, including medicine.[33] According to the Susskinds, this grand bargain is a historically embedded contractual agreement. Through it, society grants professionals exclusive rights and privileges in exchange for delivering services and maintaining high standards of expertise and integrity. As a result of this grand bargain, the professions are granted a high degree of autonomy over their practice and regulation. They also enjoy both a market and a social monopoly when it comes to status in society.

The grand bargain relies on an implicit trust on the part of the public. White-collar professions effectively control the machinery of production and distribution of their knowledge. Trust in them may not always be warranted, and there can be a pernicious, darker side to the exclusivity of this arrangement; as the Susskinds argue, the professions tightly guard their knowledge and services so that "all others (non-professionals) are excluded from becoming involved other than as recipients."[33]

In medicine, it is undeniable that such a self-protective guild mentality prevails. This is typified at times by a tendency to mystify expertise and ring-fence information, a behavior that has long been documented by sociologists.[34]

Consider the most basic form of *de*mystification—sharing access to patients' health records—an area in which I have conducted a considerable amount of research. For many years, patients across multiple countries have been afforded the legal right to request hard copies of their medical records. For example, in the United States, the Health Insurance Portability and Accountability Act of 1996 permitted patients to poke their noses in. In the UK, this statutory right was enshrined by the Access to Health Records Act of 1990, a law that was reinforced by the Data Protection Act of 1998.

Despite these theoretical rights, submitting freedom of information requests has proven profoundly cumbersome, time-consuming, and a socially awkward undertaking for patients. As we will see in this book, most patients would not dare to be seen questioning their doctor, even if they do so in private. Beyond this, under these rules there was a gray area about how much of the record could be made available, whether physicians could legally redact information and if so, under what circumstances.

For patients, rapid, online access to their own clinical information makes the process easier. Like secure internet banking, the idea is that, via online portals, patients can instantly read their own health information in real time, using their smartphones, tablets, or laptops. However, wherever patient access has been proposed or implemented in the world, it has met with vehement criticism and strong opposition from the medical profession.

This opposition is partly understandable: doctors fear patient anxiety, confusion, or increased contact in already jam-packed schedules. Yet these concerns haven't materialized. It turns out that patients can be "trusted" to read their own medical records.[35,36]

Critically, however, the routine blocking of information also carries patient safety risks that were entirely eclipsed. With access, patients better recall what was communicated during their short appointment. They better remember to follow up with referrals and test results, and do a better job remembering their treatment plan.[35-38] With access, one in five patients also report finding errors in what their doctor has documented, some of them serious.[39]

Doctors know this. For example, as part of their new NHS contract, from October 2023, general practitioners (GPs) in England are obliged to offer patients access to their medical records online. Exploring their opinions, in 2022, I led a survey which found 60 percent of GPs in England believed their patients would find "significant errors in their records," yet only one in three said online access was a "good idea."[40] In a series of stalling strategies that went back years, faced with the proposition that patients might finally read their doctors' notes, the British Medical Association—the UK's professional and trade union organization for physicians—threatened the NHS with legal action.[41]

The medical profession exerts political and economic spheres of influence. In the US it is a $4.4 trillion industry. Here professional salaries and health insurance profits versus the provision of accessible healthcare are in perpetual tension. While tech giants now wield significant political power, raising huge new ethical and social concerns, the long-standing influence of the white-collar professions has quietly shaped policies for decades, often prioritizing their own interests over those they are supposed to serve.

For example, the American Medical Association (AMA) continues to exercise its powers as one of the top ten largest lobbying groups in the US.[42] In 2024 it spent $18 million in efforts to influence government policy in Washington, comparable to the amount that Meta spent, and more than Alphabet (which owns Google) at $11 million. Meanwhile, OpenAI spent $1.76 million on government lobbying in 2024.[43]

Still, if we add to the AMA's expenditure the American Hospital Association, which advocates for the interests of current healthcare systems and hospitals, the lobbying figure is more than Meta, Alphabet, and OpenAI combined. Again, this is not to dilute the well-recognized challenges of Big Tech but merely to add that the white coats of the established medical profession are not pristine and unassailable.

Professional ring-fencing is also evidenced by monopolies on licensing and the control of other clinicians. In the US, studies show that physician associates and nurse practitioners are capable of undertaking an estimated 85 to 90 percent of the work currently performed by primary care doctors.[44] Studies also reveal that patients are no worse off when cared for by nurse practitioners than by primary care physicians, with patients often reporting similar or greater care satisfaction.[45,46]

The country's 385,000 nurse practitioners fill a much-needed gap, yet they are licensed to work without physician oversight in fewer than half of all US states.[47] Medicine as an institution is an impediment to the growth of these "second-level" professions. In 2023 the AMA doubled down, opposing the full potential of physician associates and nurse practitioners, and the establishment of autonomous regulatory boards for non-physician health professionals.[48]

Even in countries that do not operate fee-for-service healthcare, medical monopolies are clear. Today many British doctors consider themselves the custodians of the NHS and defenders of the principle of open-access healthcare for all. This is a noble cause. However, the British Medical Association vehemently resists the expansion of physician associates (PAs), whom they characterize as undermining the "unique role of the physician."[49] As of 2024, there are only 4,000 PAs working in the UK, with almost nine in ten British doctors reporting that they worry about the safety of care provided by PAs compared with them.[49] Even if we accept their concerns, the comparison in current creaking health systems is not with patients seeing PAs but—as we'll discover—with patients not accessing any care at all.

The opacity of the medical profession camouflages underperformance, patient disempowerment, and ethically questionable practices. We'll explore the extent of this underperformance and disempowerment throughout the book; however, it is worth emphasizing that questionable practices are more common than we might suppose. In 2016, improper payments connected to abuse or fraud cost the US Centers for Medicare and Medicaid Services $95 billion.[50] Examples of fraud included "up-coding," which includes physicians billing for services not performed, billing for unnecessary services, and misrepresenting the diagnosis to justify payment.

Serious ethical violations such as improper prescribing of controlled substances, sexual impropriety, and unnecessary invasive procedures are not rare.[51] A review in the US found that the annual rate of major disciplinary action in medicine resulting in revocation, suspension, or surrender of licenses was one per 1,000 doctors, which is similar to the annual incidence of breast cancer.[52] These figures may also mask a more worrying picture. In 2019 a national survey found that nearly one in five Americans experienced an interaction with a physician whom they believed was acting unethically, unprofessionally, or providing substandard care, but only one third of these patients reported the misconduct or filed a complaint.[53]

Doctors have a professional duty to report impaired or incompetent colleagues to relevant authorities but, in practice and when confronted by these situations, many fail to follow through. In a national US survey, 96 percent of physicians stated that impaired or incompetent colleagues should be reported; however, nearly half—45 percent—who encountered such a colleague admitted failing to notify state licensing boards about the individual.[54] Why? A study published in *The New England Journal of Medicine* found that "fear of how a colleague will react, along with strong cultural norms around loyalty, solidarity, and 'tattling'" deterred doctors from discussing their colleagues' errors, and identified "a natural reluctance to risk acquiring an unfavorable reputation with colleagues, disrupting relationships among and within care teams, or harming one's institution."[55]

If the medical profession has a history of inscrutability and jealously guarding its expertise, it also has a preponderance to emphasize the meaningfulness of work for its members. Medicine is often described as a calling, though it also comes with perks, including prestige and a healthy salary. Beyond these boons, many medical doctors are open about experiencing a "healing buzz." British surgeon Henry Marsh is among the most forthright about the highs; in *Do No Harm* he writes: "As I walked round the wards after an operating list with my assistants beside me and received my patients' heartfelt gratitude and that of their families, I felt like a conquering general after a great battle."[23]

Given the chronic pressures on them, it is gratifying that doctors accrue meaning and personal reward from their work. However, the function of medicine is not to preserve work for an elite stratum for whom employment carries special significance. Career meaningfulness does not constitute a justification for preserving the profession in perpetuity.

Professional prognoses

Given that the function of medicine is the care of the patient, we must ask whether doctors are essential in providing it. Many have had their doubts.

Consider this extract from a scientific publication which noted: "Computing science will probably exert its major effects by augmenting and, in some cases, largely replacing the intellectual functions of the physician.... [T]he computer as an intellectual tool can reshape the present system of healthcare, fundamentally alter the role of the physician, and profoundly change the nature of medical manpower recruitment and medical education." This proclamation comes from an article entitled "Medicine and the computer: the promise and problems of change."[56] The title could have been printed yesterday but the article was authored more than fifty years ago, in 1970, by Dr. William Schwartz.

In recent years the idea that technology could partially or entirely take over many medical tasks has gained widespread traction. Some tech gurus argue that a digital revolution is imminent and Dr. Robot or its ilk is on the way, evoking images of neurotic *Star Wars* character C-3PO scuttling toward us in scrubs and stethoscope.

In 2012, fiercely ambitious software billionaire and Founder of Sun Microsystems Vinod Khosla predicted, "By 2025 more data-driven, automated healthcare will displace up to 80 per cent of physicians' diagnostic and prescription work."[57] In 2014, Dr. Pete Diamandis, physician, and founder of the XPRIZE Foundation, told journalists: "I can imagine a day in the future where the patient walks into the hospital and the patient needs, say, cardiac surgery, and the conversation goes something like this: 'No, no, no. I do not want that human touching me. I want the robot that's done it 1,000 times perfectly.'"[58]

Also in 2014, Andrew McAfee, a research scientist at MIT and co-founder and co-director of the MIT Initiative on the Digital Economy, told reporters that IBM's "Watson"—the machine that won US gameshow *Jeopardy*—would soon be the best doctor in the world: "I'm convinced that *if it's not already the world's best diagnostician, it will be soon*" (my emphasis).[59] Caught up in the moment, respected technology magazine *Wired* declared, "IBM's Watson is better at diagnosing cancer than human doctors."[60] Infamously, in 2016, British Canadian computer scientist Geoffrey Hinton asserted, "We should stop training radiologists now. It's just completely obvious that within five years, deep learning is going to do better than radiologists."[61]

Despite the urgency and media-driven focus of these predictions, physicians are still in their jobs and patients don't routinely demand that Dr. Robot carry out their surgery. Following all the bold publicity and breathless anticipation, the five-year collaboration between IBM Watson and the MD Anderson Cancer Center eventually collapsed due to technical challenges, integration issues, and unmet expectations.

AI hasn't gone away, however, and has since picked up pace. Starting in late 2022, apps from a new generation of computer software have swiftly become ubiquitous celebrities: Dr. Bot is the most recent incarnation of the prophecy. Just as Beyoncé and Taylor Swift

PROGNOSIS AND TREATMENT OPTIONS

dominate popular culture, "generative AI" is a celebrity in the world of Big Tech.

Exemplified by OpenAI's GPT-3.5, and its later versions, GPT-4 and GPT-4o, as well as Anthropic's Claude, Google's Gemini, and the Chinese contender, DeepSeek, these tools function differently from how search engines traditionally work. Rather than typing requests for information online and receiving a list of internet links in response, generative AI chatbots are impressively interactive. Like talking to the internet on steroids, these tools hold what simulates a conversation. Because they are powered by vast swathes of data that fuel their responses, they are referred to as "large language models" (LLMs).

In 2023 global media headlines were abuzz with news about ChatGPT—the first LLM-bot on the block. Once again, prominent figures were confident that the bot and its buddies could overhaul healthcare. Venture capitalist Vinod Khosla boldly anticipated: "Within 5 to 6 years, the FDA [Food and Drug Administration] will approve a primary care app qualified to practice medicine like your primary care physician."[62] Khosla was very far from being a lone voice; diverse media outlets—ranging from *Scientific American* to *The Daily Mail* and from *The Wall Street Journal* to *Bloomberg*—ran variations of the headline: "The AI doctor will see you now."

Despite the latest wave of enthusiasm, many remain unconvinced that doctors are imminently in danger of being booted out by this new wave of chatbots. Some say that such forecasts are foolhardy. Among them, cognitive scientist, AI commentator, and author of the book *Rebooting AI* Gary Marcus argues that technology isn't ripe enough to replace anyone, never mind clinicians.[63] Because he's an AI enthusiast, Marcus is especially worth listening to.

A current cyber-cynic, he wants to see a "cognitive revolution" in AI—one that sees technology take the human mind as its muse. Although Marcus recognizes that machines do not need to be exact replicas of humans to be intelligent, and knows well that human reasoning can be flawed, he emphasizes that *Homo sapiens* offers the only model of genuine intelligence we have to go on. In his opinion, it would be deeply unwise not to deploy it as the model for more capable AI.

Others within medicine are skeptical too. Even those who actively support the use of AI tools to streamline doctors' tasks do not envisage a technological takeover. For many years, Dr. Eric Topol, who is a leading AI advocate, a cardiologist, a writer, and one of *Time* magazine's "100 Most Influential People in Health," has forecast that doctors' jobs will change, but in the long term their employment is secure. Topol says, "[AI] tools are symbiotic add-ons, not replacements for clinicians,"[64] and technology will never put the kibosh on softer aspects of care, such as empathy. Instead, a soft-focus image of "teamwork" is envisaged with AI grafted onto doctors' jobs.

Who is correct? Nobody has a crystal ball, nor are the sides in this debate neatly mapped. Notice, however, that tech investors and the medical profession have vested interests and their own agendas. Tech entrepreneurs want to sell products. They have a clear financial stake in marketing and overhyping their creations. In contrast, doctors' existential concern is with preserving their own market and social monopoly as the exclusive purveyors of medical expertise. The concerns of skeptics like Gary Marcus are different, and we will shortly return to them.

Technological treatments

A third, more nuanced, fate is forecast for doctors and, more importantly, patients. It has its roots in the thinking of the late American academic and business consultant Clayton Christensen. In his book *The Innovator's Dilemma* (1997), Christensen predicted two phases of technological transformation—"sustaining" and "disruptive."[65] The popular tech term "disruptive," frequently mentioned in AI conversations, originally comes from Christensen.

To illustrate his forecasts, the former Harvard Business School professor used the example of the movie rental business.[66] In the 1980s and 1990s people trekked to bricks-and-mortar stores such as Blockbuster to browse the film collection on a Saturday night, taking home and returning a hard-copy video or DVD, risking fines if returning it late. In April 1998, however, during a short-lived *"sustaining phase"* a new company—Netflix—started renting DVDs via mail. This opened new efficiencies in the movie rental market and offered greater convenience. Netflix was also hitting a new market: movie buffs who wanted a broader range of films than shop units could stock. The sustaining phase didn't last long. Only one year later Netflix inaugurated a *"disruption"* phase: new technologies and the internet facilitated the streaming of online films, radically changing the model of service.

The rest of the story is obvious: today when we want to watch a movie, we click a few buttons on a digital device and, for a monthly membership fee, watch as many movies as we want. Algorithms—producing suggestions collated from millions of other viewers' preferences—offer us "personalized" recommendations. The service is faster, cheaper, more convenient, and the all-round entertainment experience superior to VCRs, with (possibly) fewer family arguments about what to rent. It is also a radically different business model compared with the original concept of movie rentals.

Influenced by Christensen's insights, Richard and Daniel Susskind apply this model to the white-collar professions.[33] They argue that the professions—including but not limited to medicine, law, accounting, education, and journalism—are rapidly changing, and becoming dismantled by technology. Echoing Christensen's thinking, the Susskinds forecast two phases of technological transformation: *automation* and *innovation*. Automation (broadly, Christensen's "sustaining" technologies) refers to tools that support and enhance current ways of working. They foresee the second, innovation stage of this process (broadly, Christensen's "disruptive" phase) as carrying more radical consequences for the professions.

This book agrees with these forecasts. Technological transformation will run in parallel. In the coming years, the professions will be marked by identity crisis and flux: while many tasks become automated, others will be increasingly subject to innovation, replacing traditional ways of doing things. Innovations, in other words, will not immediately lead to the outright elimination of jobs, and old ways of working will not be immediately discarded. In the future, however, there will be a shift to different kinds of work including the need for a wide variety of new roles involving skills in AI and data science. Ultimately, doctors' jobs will be increasingly disintermediated and dismantled. In this respect, I argue, Vinod Khosla will likely be proven right. For patients, many aspects of this tech transformation could be a good thing.

Where does this leave doubters like Gary Marcus, who question the current capability of AI to replace professionals? This line of reasoning (dubbed by the Susskinds the "AI fallacy")[33] assumes the only way to perform intelligent tasks is by replicating the thinking processes of experts—the white-collar or white-coated workers. It is a commonly voiced assumption in AI medicine.

I agree that it is entirely conceivable that, in the long term, a cognitive revolution in artificial intelligence will be needed to fully replace most humans in the delivery of healthcare. Although it is early days, and extreme caution is necessary, such a revolution might already be slowly in motion. Newer models of AI—OpenAI's o1, o3, o3-mini, and DeepSeek-R1—appear to simulate more complex reasoning than LLMs like ChatGPT. More than just massive probability machines, many are touting these tools as the beginning of "agent AI," forms of technology that could soon tackle and complete challenging tasks independently. However, at the time of writing, research into these latest models is only getting underway.

Even if these tools are overhyped, many cynics are too rigidly defensive of a white-coat orthodoxy. Here are three reasons why. First, such a perspective embraces the Panglossian presumption that the status quo—how doctors currently do things—is as good as it gets, constituting an adequate measure of "success." In the pages that follow, I explain why replicating human doctors working in current health systems settles for much too low a bar for patient care. Doing so would perpetuate harm for some already marginalized populations. To improve medicine, we *should* expect to do things differently.

Second, such a view unfairly favors human physicians over technology, dismissing AI's potential without a fair assessment. We must remain agnostic about who or what can perform the function of medicine better; we need to ask challenging questions about what kinds of workers are needed. Just as Netflix disrupted the traditional movie rental business, in healthcare, roles and tasks may shift, with technology transforming existing jobs and creating new opportunities. Implicit appeals to historical antecedent provide no special pleading or enduring consolation for conserving the medical profession and its structures as they currently stand.

Third, we need to be more Baconian and less presumptuous when it comes to bots and other AI tools. Let us revisit some fundamental scientific facts and a bit of history. Francis Bacon (1561–1626) played a foundational role in the Scientific Revolution and is often called the "Father of the Scientific Method." Bacon championed a balanced approach to science, stressing the need to integrate empirical inquiry with theory rather than relying solely on either one:

> Those who have handled sciences have been either men of experiment or men of dogmas. The men of experiment are like the ant, they only collect and use; the reasoners resemble spiders, who make cobwebs out of their own substance. But the bee takes a middle course: it gathers its material from the flowers of the garden and of the field but transforms and digests it by a power of its own.[67]

Like the ants, a considerable amount of AI research blindly gathers data without a strategic understanding of the broader implications or the usefulness of its findings for real patients. Like the energetic spiders, vast volumes of column inches and airtime are devoted to anticipating and agonizing over what AI cannot do in the abstract. But as Bacon cautioned, we need to become more like bees. This means exploring the real-world value of tech tools such as bots *in practice, for real people*. Doing so will help us figure out what technology might be good for and why, including how it might complement or even surpass flesh-and-blood doctors. Moving away from the armchair, we might also be surprised by what we find. Such explorations must be rooted in the deep realities of current health systems and clinicians' capabilities, rather than idealized expectations of those capabilities.

So-called "AI winters," which follow inflated expectations about what technology can do, camouflage broader trends. Amara's Law is the observation that we tend to overestimate the effect of technology in the short run and underestimate its effects in the long term. Therefore, in the pages that follow you will see no hard timelines. In fact, by the time you read this, much of the AI content will be out of date while other hopes may be dashed. Yet it is worth remembering: the technology we have today is the worst it will ever be. And while AI has yet to disrupt healthcare in the ways that it has infiltrated other occupations and industries, as we've seen, advances are not abating. The demand and motivation have never been greater. Reflecting social needs, the AI healthcare market is projected to grow from $21 billion in 2024 to $148 billion by 2029—an annual rate of around 50 percent.[68]

Doctors are already using bots like ChatGPT in their practice. In October 2023, I led the first survey investigating psychiatrists' use of commercial generative AI bots.[69] The sample size was small—138 doctors affiliated with the American Psychiatric Association: four in ten said they had used ChatGPT-3.5 "to assist with answering clinical questions."

Digging deeper, by February 2024, a year after ChatGPT was released, I conducted another survey—the biggest of its kind—examining doctors' use of LLM chatbots. This time my team surveyed more than a thousand GPs across the whole of the UK to investigate the uptake of these tools in primary care.[70] In total, one in five told us they were using these tools to assist with clinical practice, including for diagnostics and treatment recommendations. This study received considerable media attention. But in public debates with me, some doctors clung to their skepticism, unwilling to fully admit or accept what their colleagues were surreptitiously doing. Yet, by January 2025, in a re-run of the survey, the increase was clear: one in four UK GPs now told us they'd adopted these tools to assist them.

Patients are also increasingly leaning on the silicon shoulders of bots for healthcare advice, turning to them for diagnostics, treatment information, and even emotional reassurance. This book will explore how, and why.

Depending on one's stance, the idea of outsourcing medicine to machines or even de-doctoring medicine may sound dystopian, utopian, or plain far-fetched. The belief that we live in the best of all possible worlds takes the path of least resistance. It is tempting to romanticize human capabilities, especially those of our current doctors whom we know, respect, and admire.

If technology is to improve on the traditional doctor–patient appointment, it will demand genuine effort and a clear focus on why it truly matters. That is the task of the rest of this book. So let us delay no further and meet our first patient—Jen.

PART I
ACCESSING CARE

2

Patient Pilgrims

"If one more clinician asks me, 'Have you tried yoga?'" says Jennifer Lawson, "I feel like I could punch them. Seriously. 'Have you tried meditation?' Fuck you."

Jennifer is not a violent person. She never loses her cool. In fact, she is physically incapable of punching anyone. Gentle by disposition, her patience has been stretched to breaking point when it comes to her long-term health conditions. Jen—as she is known to family and friends—is now 50 years old, and since the age of 3 she has lived with the hereditary condition Ehlers-Danlos syndrome or "EDS," affecting the body's connective tissues which play a crucial role supporting the skin, bones, blood vessels, and many other organs. EDS has a wide range of presenting symptoms, from joint pain to chronic and even life-threatening complications. It is the same condition portrayed in the British soap *Coronation Street*, where the character Izzy Armstrong, played by actress Cherylee Houston, reflects her own real-life experience with EDS.

Although Jen Lawson usually wears her experiences lightly, with an acute sense of the absurdity of life and of luck, she is blunt about the lived reality of her illness. EDS has created a cascade of related ailments, and they have worsened in recent years.

"I had a real breakdown last June. Stuff was going sideways left and right. I wasn't getting continuous care. I wasn't getting the tests, the laboratory and diagnostic tests that I needed."

"I was suicidal," she says. "I feel like I don't matter in this medical system. For people like me it's a broken system. I'll wait to be referred, and then see a specialist for ten minutes and they're like, 'Okay, see you in a few months' time.' To this day, I feel like they're just waiting for me to die."

Jen's health

Jen talked with me from her home where, due to failed spinal cord stimulator surgery for her EDS, strategically placed pillows on her sofa and bed are one of the few ways she can prevent further injury and cope with the now-relentless pain.

"When I was younger, dislocations were constant—by that I mean daily. Ankles, elbows, shoulders." From an early age, everyday activities posed problems: "I had a really hard time holding a pencil as a kid because I didn't have enough muscle tone to grasp it properly. The whole finger would just slide out of joint. I also fell a lot. I was just a really clumsy kid."

Awkward limbs and searing pain were always part of Jen's life. But even during those early years, she strove to normalize her differences. "I did party tricks in school. I'd pop both my shoulders out to freak out friends. *Boy was it painful*. But, yeah, I learned not to show it. I made fun of it." Jen no longer mocks her dislocations but discusses her medical challenges with the same generous, easy rapport I imagine she shared with her classmates.

Like many patients with chronic illnesses, she has both intimate first-hand knowledge and considerable expertise from decades of reading about, and keeping up to date on, research into her condition. EDS affects at least 1 in every 5,000 people. For more than four decades, Jen has experienced back pain, and over the last thirty years, since her late teens, she has endured mobility problems with her knees. "My pain is now so intense," she tells me, "some days I have to lie in bed for a full 20 hours. Nobody chooses to live like this."

"Many scientists looking into EDS find overlaps with other disorders," Jen says. "And since so many of these conditions tend to cluster, some suggest we might one day discover a global disorder that isn't necessarily EDS but something broader."

Jen explains, "For example, one related condition I have is dysautonomia, which is basically dysregulation of the autonomic nervous system. It messes with everything from heart rate, to pain, and even my digestive tract."

"My doctors think I may have a kind of dysautonomia called POTS [Postural orthostatic tachycardia syndrome], meaning I have very low blood pressure but a high heart rate. When I stand up I sometimes black out because of the drop in blood pressure. I've experienced this since I turned 10 or 11." Since her teens, Jen stands at a model-esque 6 ft tall, so falls are hard.

EDS is also commonly connected with another set of conditions called "mast cell activation diseases." She explains: "Mast cells are a type of cell that causes inflammation and is part of the system that causes a histamine response in your body—when your immune system is fighting allergens."

"Lots of people who have mast cell diseases experience allergies. *And hello, that's me.* I'm allergic to ridiculous things like aloe, gluten, dairy, green peppers, lots of food allergies, and all sorts of medications."

Jen's allergies mean the pain associated with EDS is especially difficult to manage. "I'm allergic to every single NSAID [a class of drugs that reduce pain and inflammation] out there. I can't take a single one of them on the planet, nor a steroid called methylprednisolone, which is the one they used for COVID treatment. They also use it for allergic reaction to bee stings, all sorts of things."

"By the way," she says, "I'm aware I might sound like a medical moaner. But hey, that's just the *glamorous* life of living with EDS."

Despite all her health problems, and my line of questioning, Jen isn't prone to self-pity. When she talks, she might as well be wearing a white coat, such is her critical distance and infusion line in gallows humor. She's equally upfront about the psychological toll illness has wrought: "Depression and anxiety are biggies."

Jen offers an example: "A lot of people with EDS have digestion problems. I've always had a *really* dodgy stomach and issues with constipation and diarrhea. Since I've been a kid, I've experienced constant social anxiety because you're thinking, '*Can I leave the room if I feel bad?*' Or '*Is there a bathroom nearby?*' It's often an emergency situation. That kind of thing can really fray a person's nerves."

In 2020, Jen's pain was so relentless, she contemplated suicide: "From a scale of 1 to 100, I'd say my pain was about an 85. It was burning so bad that I wanted to cut my arm off. I was not sleeping. I was literally grabbing my leg, causing bruises it was so bad."

It is no exaggeration that EDS has impacted every aspect of Jen's life. Yet, despite everything, she says, "I know that I'm one of the lucky ones."

Before we even set foot in the doctor's office, like Jen Lawson, our health is determined by an entanglement of factors. Genetics, where we live, economic circumstances, our education, our social status in society, the language we speak, and our personal support networks intricately combine to influence whether, and how, we see doctors at all.

Our journey through the traditional medical appointment starts with how we access the doctor. What we'll discover is alarming. Healthcare is topsy-turvy. The people who need it most—like Jen—often struggle the hardest.

A gal from Kalamazoo

"Yes, there really is a Kalamazoo," read slogans sold on T-shirts in the city and county of Jen Lawson's birthplace. The small Midwest city is nested on the southwest bank of a bend in the Kalamazoo River between Detroit and Chicago. When Europeans first arrived in the seventeenth century, it was the Potawatomi people who called this place home. More than 400 years later, the city has 75,000 inhabitants.

Over the years, K'zoo—as it's called by locals—has undergone multiple reincarnations as a trade and manufacturing base. Once dubbed the "Celery Capital" of the US, in the first half of the twentieth century it became known as "Paper City" and was renowned nationally for its paper-making mills, but when the forests were logged the industry closed. It was the 1940s when Glenn Miller and His Orchestra put the place on the map with the hit song "(I've Got a Gal in) Kalamazoo." After that the city plucked its way into rock 'n' roll history: Gibson guitars were crafted here for generations. Another Americana symbol, the Checker Cab Manufacturing Company's headquarters were based in the city until production lines ended in the 1980s. In 2015, the long-established Kalamazoo pharmaceutical company Upjohn was bought by Pfizer of COVID vaccine acclaim.

"We've had our ups and downs. A lot of really rich families were based in this town," Jen tells me, "and because of their philanthropy there was always a lot of money for the arts, music, and theatre, including educational programs for schools. Kalamazoo was always an artsy place," she says, casually adding, "Back in the '90s, a family friend wrote a play on my experiences with Ehlers-Danlos syndrome called 'Stick Girl.' It was performed locally. Mom and Dad really liked it, but my extended family were horrified." She adds with a chuckle, "*Man, they hated it.*"

Despite K'zoo's manufacturing dynamism and creative spirit, Jen has observed changes: "Today, '*the haves*' have more and the '*have-nots*' have way less. Downtown Kalamazoo is empty. Even prior to the pandemic beautiful apartments were lying unoccupied. No one can afford the rent. It's a ghost town."

According to the US Bureau of Labor Statistics, in August 2024, around 4 percent of Kalamazoons were unemployed, mirroring the national average.[1] Jen tells me about her own employment history in the area.

"*I've always been a worker*," she says, resolute. "At 16 my first job was at the public library in the basement, reordering the books on the cart. Then I worked at a downtown pizza joint. I've even been a medical transcriptionist. And for nearly twenty years I worked at Western [Western Michigan University]. During most of that time I was a disability services coordinator." Because of her own experiences on campus, the job came naturally.

By 2017, however, Jen was ready for a new challenge, landing her "dream job"—technical writing for a mortgage company in their policies and procedures department. She adds, "I'm aware this makes me sound like a weirdo. But I love the precision behind this kind of work."

"In truth, it was a nice breath of fresh air not to have to think about disability 24-7. With the new job, I started doing exactly what I had wanted to do for a long time."

Jen's dream job was cut short. "I had a very serious fall in my house. My legs gave way and I fell down two stairs and dislocated my shoulder and my knee, and subluxed [partially dislocated] my left wrist. With EDS, a fall is not a normal fall. That's on top of all the daily issues that go along with living with this illness. I couldn't use either wrist. The knock also started a shitstorm of stuff with my body."

Jen says: "All of a sudden it was *boom*, not working. Luckily the company offered me short-term disability for six months. After that point they said, 'Bye.'"

According to the World Health Organization, falls are the second leading cause of accidental injury deaths worldwide. Adults older than 60 years of age suffer the greatest number of fatal falls,[2] and people with physical disabilities face the biggest risk of accidental injury.[3]

"I had a wake-up call: my body was wrecked from working too hard for too long, and just pushing through." Unlike the young Jen who performed party tricks for her peers, adult Jen knew she could no longer normalize her illness.

"Luckily," she tells me, "under the Affordable Care Act that Obama passed, they expanded coverage for people like me. Before I was eligible, I had health coverage under COBRA, which is a stop-gap measure—in between jobs you pay for your health insurance to continue under your employer. I did that for 18 months after my job ended, that was around $600 a month. I blew $80,000 on healthcare, paying it out of my retirement account because I didn't have any income coming in."

Despite living with her elderly parents to save money, she has now used up all her hard-earned savings on medical costs. Jen is hoping her application for Medicare—the federal government insurance program for people over 65 and those with disability needs—will soon be approved.

"When my funds ran out, I went onto state-run Medicaid. I have no money, and I'm basically indigent. So, I qualify."

Wealth is health

Worldwide, nearly 100 million people each year are propelled into extreme poverty by the costs of healthcare.[4] A study of 153 countries in The Lancet reported that "broader health coverage generally leads to better access to necessary care and improved population health, with the largest gains accruing to poorer people."[5]

The US is unique among the 33 wealthiest Western countries in failing to guarantee universal care to all of its citizens.[6] Inability to pay for coverage carries considerable consequences for health-seeking behavior: personal purse strings influence how—or indeed, whether—people access clinicians at all.

More than 30 million Americans—nearly 10 percent—have no insurance.[7] Out-of-pocket medical costs and being uninsured in America increase the risk of death by 17 percent.[8] Despite the

Affordable Care Act, medical bills still contribute to over 60 percent of bankruptcies.[9] During the COVID-19 pandemic, unemployment surged to 20 million, with an estimated 40 percent losing employer-based health insurance through lost jobs.[10] Researchers at Yale School of Public Health concluded, "a single-payer universal healthcare system would have saved about 212,000 lives in 2020 alone," and "US$105.6 billion of medical expenses associated with COVID-19 hospitalization could have been averted by a single-payer universal healthcare system over the course of the pandemic."[11]

Most countries in the European Union (EU) offer universal access to at least a core set of health services. A joint Organisation for Economic Co-operation and Development (OECD) and EU report concluded that only a small share of people experienced unmet needs for healthcare.[12] However, even this comparatively roseate picture masks serious challenges that affect only some populations. Around a fifth of healthcare spending in Europe comes directly from patients' pockets, and across the whole of the EU, the proportion of people with unmet medical needs is still nearly five times higher in poorer households. For low-income groups, the OECD–EU report notes these payments can be "catastrophic," tipping people into poverty.

Jen Lawson's health places her in a seldom highlighted predicament: disability poverty. Today, around round one in seven of us worldwide live with some form of disability,[13] which can inflict literal costs. Health insurance and life insurance are higher, and opportunities for work are tougher. In 2024, the US Bureau of Labor Statistics reported that 7.2 percent of people with disabilities were out of work, more than twice the rate as for able-bodied Americans.[14]

Research is more limited than for other patient populations but, in 2015, a study in the *American Journal of Public Health* reported that, in the past year, more than a quarter of people with disabilities who needed to see a doctor skipped or avoided doing so because of costs, compared with 12 percent of those without disabilities.[15] Investigators found that even with health insurance, 16 percent of people with disabilities forwent care due to expenses, compared to 6 percent of those without disabilities. Even when they can afford medicine or have coverage, those with disabilities face additional living costs, such as home adaptations, assistive devices, and personal assistance.[16]

Like a quarter of all Kalamazoo's residents, Jen has Medicaid coverage. But she has recently confronted another healthcare home truth: she isn't a priority patient.

"Doctors don't get paid enough to see people like me. There are a lot of doctors who just don't accept my health insurance. I'm stuck between a rock and a hard place because I can only get what they give."

In the US, Jen's illness affected her employability, which affected her income which, in turn, affected her health insurance—a chain of events that exacerbated the stress of living with chronic illness. In a lose-lose situation, epidemiologists point out that the perception of control that we have over our lives can exert biological effects on our body, kick-starting a cascade of stress on our autonomic nervous system.[17] Elevation in heart rate and blood pressure, and spikes in cortisol levels can exact a serious physiological toll, storing up long-term damage by increasing the risks of heart disease and mental health problems.

Medical mileage

Obtaining health coverage, or living in a country with universal healthcare, is a good start. But it is only the beginning of the assault course people with chronic illness must negotiate to obtain care.

"Another problem with EDS," Jen adds with resignation, "is there aren't many doctors nearby who can treat me."

Kalamazoo may have been the Mecca for guitar and taxicab manufacturing, but it was never the medical epicenter of America, let alone Michigan. Boston, Massachusetts, in contrast, is home to 25 major hospitals. When it comes to bricks-and-mortar clinics and hospitals, luck and the caprices of history play a part.

ACCESSING CARE

Inevitably, some of us will face extended journeys to medical facilities. A fifth of Americans live in rural communities where healthcare is less available, facing longer travel times, and greater risk of losing local services.[18] Approximately 60 percent of "Primary Medical Health Professional Shortage Areas" are based in rural areas.[19] Across EU member states, those living in rural areas report greater access problems than their counterparts in urban areas.[20] In 2018, an EU report noted that approximately six in ten people living in villages could drive to a main healthcare facility within 15 minutes. Fewer than half of those living in mostly uninhabited areas could do the same.[20]

In the US, zip codes combine with health coverage to create additional barriers. On Medicaid, patients cannot go out of state to receive medical care. However, after a few months, Jen's family doctor finally referred her to a specialist 100 miles across state in Ann Arbor.

"At the time, I was so grateful," says Jen. "Except that it's a two-and-a-half-hour drive. I can't drive any longer than about 35 minutes at this point—even that's a struggle. Mom is 80. She can drive longer, but she's not good on busy highways. So, for me it was like, 'Great! A specialist at last—awesome! But how am I going to get there?'"

Jen assiduously attends all her appointments but relies on her elderly mother. She lies in the back of the car with cushions and pillows to ensure she doesn't get injured. In many countries and health systems, patients in need are entitled to financial reimbursement, usually for the cheapest suitable mode of transport. Jen says, "Oh sure, I qualify for assistance. But it always means taking longer journeys with detours and discomfort. It turns everything into an even bigger pain in the neck—literally."

The reality, in turn, involves more hefty costs. "Traveling to the specialist means a full tank of gas round trip for us," says Jen, who tells me this wasn't the only hidden expense. "The last time we went up to Ann Arbor, we had to go up the night before and stay over in a hotel because it's too much to do round trip in one day for me. Hotels there are not cheap. We spent $174 for a room. That was the cheapest we could find close to the hospital. How are people like me supposed to do that when we don't have any money?"

Transportation costs highlight that lack of universal health insurance is a major obstacle to care, but not the only one.[21] For example, in 2006 in Massachusetts state authorities embarked on reforms to expand health coverage among the underprivileged. Two years later, there was progress, but around four in ten people who had been fully insured for the previous year still faced at least one other hurdle in the assault course to care.[22] Nationwide, two out of three Americans experiencing financial barriers cite at least one other impediment to access. Barriers are more common among younger people, women, parents, those on lower incomes, and those with at least one chronic illness.[21]

The US has one of the worst public transit systems in the developed world.[23] Transportation problems are a common reason why people fail to attend medical appointments,[24] a barrier that predominantly affects people with low incomes. For those who don't own a car or who live without a direct public transport link, attending health checkups can seem like an impractical burden or logistical nightmare. Nor are these burdens confined to rural communities. In a study of residents of a New York suburb, people who relied on buses were twice as likely to miss medical appointments as those who drove.[25]

Consider disability again: one US analysis found that 34 percent of people with disabilities, compared with 16 percent of people without disabilities, reported inadequate transportation to access care.[15] In the UK in 2017, a study found people with disability have two to four times higher odds of experiencing unmet healthcare needs due to transportation barriers.[26] In the EU, the level of unmet medical needs is two and a half times higher for people with disabilities than the overall population. A recent EU report placed "exorbitant treatment costs, inaccessible transport and health facilities" at the top of a list of barriers for people with disabilities.[27]

Time is money: among those holding down multiple gig-economy jobs, attending medical appointments, including health screenings, creates additional logistical and financial pressures. American Time Survey Data gathered from 2005 to 2013 found, on average, patients were required to take more than two hours out of their day for a 20-minute doctor's visit.[28] Patients who were unemployed or on low incomes experienced a 25 to 28 percent greater time burden in accessing medical visits compared with wealthier populations. Parents and those in part-time work also report more scheduling and availability problems in attending medical appointments. Perhaps as a result, in the US, more women than men—around one in four women compared with one in five men—cite travel as a health appointment barrier. Survey data from 70 countries reveals that in nations with higher gender inequality, women experience greater mobility challenges in accessing clinics and hospitals.[29]

Like Jen, even if they finally make it to the appointment, many patients and their carers who are traveling by car—often by necessity—are confronted with another hidden tax. Hospital parking fees can be prohibitive—costs rarely reimbursed by US insurance companies. In 2020, a review in the US found that breast cancer patients paid up to $850 in parking fees during their course of treatment, while patients with leukemia forked out double this amount at up to $1,680.[30]

Jen's journey to the appointment doesn't stop when she gets there. When she arrives the physical exertion really begins.

Bricks-and-mortar barriers

"Medical clinics and hospitals weren't designed for chronically ill people like me. Ironic, isn't it?"

Jen explains: "A perfect example is a doctor I see in town. To get to the building, even with accessible parking spots, it's probably 20 meters. Then you have to walk up a ramp or upstairs. Then you're talking another 100-meter walk to the elevator, and then down two flights of stairs, and probably another 20-meter walk to the office."

Of the quarter of the adult population in the US with a disability, around one in eight experience mobility disabilities such as serious difficulty walking or climbing stairs.[31] Mobility limitations disproportionately impact the elderly, with 35 percent of people aged 70 or older affected.[32]

In case there is any room for doubt, Jen quietly asserts: "Navigating long walks in hospitals is *huge*." For her, the inadequacies of clinic adaptations pose medical hazards. "Even if I had a wheelchair, I can't push myself. *Can't do it*. Not with my right arm—not unless I'm to arrive injured."

"Besides, not all patients can afford a wheelchair, or a scooter, or someone to push us. And really, do I always have to bring someone to my appointment so they can push the hospital wheelchair? Because there's no one ever available at the hospital to help. I'm sorry. They say there is but that's bullshit. There's never anyone available to come push you."

The passage of landmark legislation has helped people with disabilities, but it hasn't stretched far enough. For example, the Americans with Disabilities Act in 1990 established the right of equal access to public services offered by governments and private providers, while the UK's Equality Act of 2010, which outlaws discrimination arising from disability, and a raft of policies and legislation enacted by the EU in the 2000s have striven to create equal opportunities for people with disabilities and prevent discrimination. Yet substantial barriers persist within healthcare facilities.

Arriving in the waiting room, "The seating is just terrible," says Jen. Discomfort is not aided by delays in being seen. In the US in 2017, an analysis of 21 million outpatient consultations found almost one fifth of patients endured delays longer than 20 minutes.[33] Medicaid patients were 20 percent more likely than the privately insured to sit more than 20 minutes.

"I try to sit like a statue and just cope. Because if I start shifting around, I almost invariably sublux a vertebra. So, the more I squirm, the worse it gets. Sitting as still as possible, deep breathing, closing my eyes, and just waiting works best."

Jen adds, "But I think this works against me. When you deal with it by sitting silently and trying to concentrate, people think you're just fine."

Entering the doctor's office there are more physical barriers. Jen's tone shifts from resignation to ridicule: "You know what fascinates me? Doctors always sit on cushy chairs. As a patient, I've never been offered the comfortable seat."

Since she hit 40, Jen says, she is more forthright, but prudent: "The younger me would have been worried about bothering them. Well, I do care about bothering them because I have my life in their hands, but in another way, I've less to lose."

"The minute they're out of the room, if there's one of those beds, I lie down on it and put the pillow underneath my knees." This resourcefulness isn't always welcomed: "Last year, a pain management doctor asked, 'What are you *doing* on there?' I said, 'Sorry, those chairs are *not comfortable*.'"

Problems extend to diagnostic equipment too. "Over the last two to three years, I've had quite a few MRIs and CT scans. I do understand that positioning someone with my condition is really hard, requiring extra pillows and so on. But I know it can be done because one time it was done well. Two techs were willing to work with me to help get me positioned right."

A commentary in the *Journal of the American Medical Association* emphasized, "Fractures, nerve and soft tissue injuries, and shoulder subluxions can result from unsafe transfer techniques."[34] In another secret-shopper-style study, investigators explored provider reactions to a fictional wheelchair user by phoning more than 250 specialist medical practices in four US cities. Upon disclosing the patient could not self-transfer from a chair to an examination table, a fifth of practices responded that they could not accommodate the caller, and a similar proportion said they would be unable to lift the patient; 4 percent said outright that their building was inaccessible.[35] In 2021 a

US survey found only four in ten physicians were "very confident" about their ability to provide the same quality of care for patients with disability, and one fifth "strongly" agreed that the health system treats people with disability "unfairly."[36]

Like Jen, patients also report attitudinal barriers. The World Health Organization estimates that, when accessing the same medical services, people with disabilities are around four times more likely to report being treated badly, and three times more likely to be denied care than able-bodied patients.[13] In 2023 the EU disability report *Denied the Right to Health* identified stigma but also "a healthcare workforce that lacks proper training"[27] as top disability barriers. In 2019, a UK survey of patients with disability accessing cancer care found patients' "needs were recognized but ignored." Upon asking for accommodations, one British interviewee—Gavin—recalled being told, "It's a lovely mattress, you stop complaining."[37]

Upside-down doctor access

As Jen's journey demonstrates, access to doctors is upside down. The healthier you are, the more privileged you become, and the easier it is to receive medical attention—a problem that has been recognized for years. In 1971, writing in *The Lancet*, British GP Julian Tudor Hart called it "the inverse care law," and observed: "The availability of good medical care tends to vary inversely with the need for it in the population served."[38]

Although Tudor Hart's concern was primarily poverty, his law also applies to a wide range of would-be patients: those with disabilities, mental health conditions, the elderly, minorities, and people living in rural areas, all of whom—for a variety of different reasons—experience a higher burden of disease yet are more likely to face challenges to seeing clinicians. Consider the coronavirus pandemic: in the US in 2021, a federal report concluded that the heavy toll of COVID-19 fatalities among people with disabilities was "foreseeable and disproportionate."[39]

In England, a report by the Office of National Statistics revealed that from January to November 2020, adults with disabilities, who make up 17 percent of the population, accounted for 59 percent of COVID-19 deaths.[40] We'll explore disability discrimination further, later in the book.

Jen acknowledges that she is luckier than some. Her ancestry is mostly European and this may present a health boon. In the US, an immense amount of research shows that Black Americans endure poorer health outcomes compared with whites for nearly every listed illness and disease.[41] Starting to address the problem in 2002, the Institute of Medicine published a landmark report entitled *Unequal Treatment: Confronting Racial and Ethnic Disparities in Health Care*. The report concluded that "Evidence of racial and ethnic disparities in healthcare is, with few exceptions, remarkably consistent across a range of illnesses and healthcare services."[42] There has been progress: between 1999 and 2015, the US Center for Disease Control and Prevention found that, for all causes of death, the racial disparity gap dropped from 33 percent in 1999 to 16 percent.[43]

However, by 2024, a *Lancet* study found US racial and geographic health disparities had widened over the previous two decades, exacerbated by the COVID-19 pandemic, with growing life expectancy gaps.[44] For example, Black Americans contracted COVID-19 at three times the rate of white Americans, and in some places the morbidity rate of the Black population was five to seven times greater than of the white population.[45]

Understanding the intricate relationship between race, ethnicity, socio-economic factors, and health requires us to unravel deeply intertwined threads—a meticulous and complex process.[46] In the American health context, a sense of history is crucial and, while immersion in the past is not always constructive or beneficial, blotting it out would be parochial.

We needn't crane our necks to look too far back to observe how racist policies enacted by US institutions and legislature contributed to healthcare disparities that persist today. The American Medical Association (AMA)—the largest association of doctors and medical students in the US—was founded in 1847 but, for more than a century, it excluded Black physicians, who were also barred from many state and local medical societies. Black doctors were obliged to establish their own medical schools and their own professional organization, the National Medical Association. Only in 2008, fifty years after Blacks had begun to be incrementally admitted, did the AMA formally apologize for its discrimination policies.

In living memory, federal law was in cahoots with the AMA to enshrine segregation in US healthcare. Under the "separate but equal" Jim Crow laws of the late nineteenth and early twentieth century, patients were treated in wards or even hospitals reserved for "whites only." After a series of civil suits, the crucial turning point came with the Civil Rights Act of 1964, which banned discrimination based on race, color, or national origin for any US agency receiving state or federal funding. However, centuries of racism and "redlining"—whereby federal government and lenders would literally draw a red line on a map around the neighborhoods they would not invest in—also left a legacy on the social and built environment. Consequently, migration patterns—both forced and voluntary—contributed to geographical segregation along racial lines.

Take Kalamazoo: while many neighborhoods are mixed, it is still possible to map racial demographics. The city's two hospitals are situated in the north and downtown areas, which are closer to predominantly Black areas of town. Both hospitals house primary care departments but most of the large, dedicated family medicine practices are located further out from the city center in the suburbs. These clinics are in predominantly white areas, imposing further travel burdens and costs on the Black population.

Throughout the United States, history has wrought differential access to health services. In 2022, a Pew Research survey found that

85 percent of Black adults believed one reason for worse health outcomes among Black Americans is "less access to quality medical care where they live."[47] After controlling for income, African Americans do indeed face a greater burden of travel to obtain medical care compared with their white counterparts.[24] The odds of any given area having a shortage of primary care doctors is nearly 70 percent higher for majority African American neighborhoods.[48]

Even relatively recent policy decisions negatively affected Black communities. Between 1937 and 1980, one analysis found hospitals primarily serving non-white populations were more likely than neighboring facilities to have been shut down.[49] Nearly 40 years later, in 2022, an examination of access to healthcare found rural zip codes with high percentages of Black or American Indian/Alaska Native residents were located significantly further from hospitals offering emergency services, trauma care, obstetrics, outpatient surgery, intensive care, and cardiac care.[50]

Jen's journey shows a variety of forces can combine to create sometimes subtle, sometimes seismic, differences in our health. Already disadvantaged populations must surmount formidable barriers to visit their doctors. As patients, it might seem, the really terrible thing is we must just accept the system.

"By the way," Jen tells me, "after our road-trip to Ann Arbor, the specialist came in, and she kind of looked at me and said, 'Yeah, okay, you're going to need to see a bunch of other people.'"

Despite lying in the back of a car to travel across the state, relying on her elderly mother to drive, and spending money on travel and overnight accommodation, the medical visit was over in minutes. "I was like, are you effing kidding me? I couldn't believe it. She hadn't even touched me."

It would be another three months before Jen Lawson would see a specialist.

3

Web-side Visits

Jen Lawson's epic journey across the great state of Michigan was agonizing. For Jen such visits reeked of physician frivolity. Suddenly, in the spring of 2020, a mindless virus axed her medical road trips.

"The minute COVID hit I saw my doctors in my bedroom on my iMac. My provider called and said, 'Do you want to do a telemedicine?' and I was like, '*Good!* This will be *so* much easier.'"

The term "telemedicine," which is sometimes used interchangeably with "telehealth," encompasses a wide variety of telecommunication services including the use of medical tests and remote monitoring of symptoms—all delivered without the need for an in-person appointment. One type of telemedicine is the use of video or telephone calls to conduct appointments with doctors.

During the nadir of COVID-19—when clinics slammed shut their doors, and governments enforced stay-at-home measures to prevent viral spread—the sweatpants-wearing public first became acquainted with telemedicine. The concept, however, wasn't new; the medium is at least as old as TV, with experiments dating back to the 1930s.[1] The first real-world application was in 1967, on the passenger concourse of Logan Airport, Boston. Though short-lived, the aim was to facilitate medical emergencies arising among the 12,000 airport employees and the 50,000 passengers who passed through Logan daily.[2] "Virtual visits" was an idea born before its time, and soon after its Boston debut, telemedicine was largely shelved.

By the late 1990s, a constellation of factors aligned to resurrect video visits as a medium with serious potential. Powered by the home computer revolution, escalating internet use, and decreasing equipment costs, the world was teeing up for telemedicine—in theory at least.

In 1995, a commentary in *The Lancet* entitled "Telemedicine: still waiting for users" noted that, despite a "surge of interest in the internet," the innovation was painfully slow to take off.[1] While the dot.com bubble frothed, and internet banking became baked into our lives, telemedicine remained an otherworldly pursuit—quite literally. In 1997 NASA deployed the medium to treat the medical needs of astronauts in space.[3] Three years later, more than 40 percent of US households had an internet connection[4] but among the 800,000 practicing doctors in the United States, fewer than 0.5 percent offered video-based visits.[5]

If war is the locomotive of history, the battle against coronavirus was the catalyst for the implementation of telemedicine. The medium got its big break and by mid-2020, uptake was dizzying:[6] across American primary care, remote visits skyrocketed from fewer than 2 percent of all appointments in 2019 to 35 percent by April 2020.[7] Or take Ireland: by March 2020, the use of telemedicine increased by fivefold, with 20 percent of the population using it.[8]

In this chapter we'll explore why it took a mindless virus for this to happen, and whether the technology was worth the wait. We'll also explore if, for patients like Jen, phone and video visits can truly flip the script on inequities in healthcare access.

Healthcare with home comforts

In 2020, one retired GP writing in the *BMJ* declared that telemedicine "violates human dignity by reducing people to the moral status of objects."[9] Equally dramatic, in 2015, in *The New England Journal of Medicine* another doctor compared telemedicine to an online dating app: "Tinder makes dating quicker, efficient, and more accessible," he wrote, "but is it better?"[10] Swiping right might not be the ideal analogy for patient–doctor relations, but it is reasonable to wonder: do people want telemedicine?

When you ask patients themselves—*mirabile dictu*—you get a rather different take on matters. "It's easy to overlook," Jen confessed over video link, "but being at home, at least for me, it's a very big positive."

"When my doctors or therapist talk with me from my bedroom, I'm already here, ready to turn it on. By the time we start, it's a much more relaxed '*Hi, how's it going?*' I don't have that anticipatory anxiety of: '*oh my God, am I going to get there on time?*'"

Jen adds with a burst of enthusiasm, "And it has definitely changed the timbre of my therapy sessions. By the way, I've never met my therapist in real life, but I've never felt weird about that relationship either."

As the world went into lockdown, surveys in the US increasingly reported "patient satisfaction with video visits is high" and "not a barrier toward a paradigm shift away from traditional in-person clinic visits."[11] By December 2020, a study of 1.3 million Americans confirmed it was the "intimacy of virtual encounters" people appreciated.[12] In September 2021, researchers at the Harvard T.H. Chan School of Public Health found eight in ten respondents were satisfied with telemedicine.[13] Unsurprisingly, those receiving end-of-life care said video links from home were preferable to being summoned to sterile waiting rooms or clinic offices.[14–16] As one patient summed it up: "Technology is an absolute godsend."[16]

Jen suffers from chronic pain, but every patient's needs differ. In June 2020, a survey by Doctor.com in the US reported that one in three respondents expressed a preference for telemedicine, with the remainder favoring in-person appointments.[17] Despite this, more than eight in ten said they'd be willing to use the medium after the pandemic was over.[17]

Lack of choice in how one accesses care colors attitudes. Telemedicine is no cure-all for healthcare's staffing shortages: it is still constrained by the number of doctors who can take calls.

For example, in the last decade, the UK has endured a lower level of capital investment in healthcare compared with European Union (EU) countries,[18] subsisting with lower numbers of doctors per head than on the continent.[19] Underfunded and understaffed, the National Health Service (NHS) has struggled to offer patients timely appointments. With the switch to telemedicine during and after COVID-19,[20] many British patients perceived the medium—which, in the UK, is frequently a phone call rather than a video visit—as foisted on them to cope with costs.[21] Some associated telemedicine with "lazy GPs."[22] Despite these concerns, patients perceived telemedicine as perfectly acceptable for some medical problems and, like their American counterparts, eight in ten said they would use it again.[23]

Telemedicine was also clearly amplifying access. During the pandemic, as demand for mental health services surged, teletherapy reached record levels.[24] In monthly growth during 2020, Teledoc—the oldest telemedicine company in the US—reported video-based mental health visits with men increased faster than those with women, and that older people were coming on board too in hitherto unseen numbers.[24] Teledoc also experienced a doubling of its services among patients who accessed care with Medicaid.

In the first year of COVID-19, the United States saw over 99,000 drug overdose deaths, an increase of almost 30 percent from the previous year.[25] However, researchers at the US Centers for Disease Control and Prevention, the Centers for Medicare and Medicaid Services, and the National Institute on Drug Abuse found that patients with opioid use disorders who used telemedicine did a better job of sticking to their treatment plans.[26] The team also identified a 35-fold increase in uptake of addiction services when people were offered the medium. Telemedicine was not just treating patients, it better identified clinical needs and helped to save lives.

Evidence gathered before COVID-19 strongly anticipated these findings. In 2016, a literature review in mental healthcare concluded there was "strong and consistent evidence of the feasibility of this modality of care and its acceptance by its intended users."[27] Multiple studies reported that telemedicine improved attendance rates and adherence to psychotropic prescriptions, and increased patient and clinician appointment satisfaction.[28,29] A randomized controlled study at the US Department of Veterans Affairs (VA)—the nationwide healthcare system offering 9 million US army vets government-funded healthcare—reported patients with mental health diagnoses who were given digital devices set up for video visits were 20 percent less likely to drop out of care six months later.[30]

In 2020, during the first year of the pandemic, more than 28 million people with Medicare—the US government health coverage program for seniors and those with disabilities—accessed telemedicine, an 88-fold increase from the year before.[31] Remote consultations helped halt the transmission of COVID. Telemedicine also helped people with mobility challenges to avoid tackling an assault course to reach the holy grail of healthcare.[32]

Reducing the ramps

Dr. Lisa Iezzoni knows first-hand the travails of traditional office-based appointments. During her first year as a student at Harvard Medical School, Lisa was diagnosed with multiple sclerosis (MS). Qualifying as a doctor in 1984, she was never permitted to practice as a licensed physician because of her condition. Instead, combining her clinical training with first-hand experiences of living with MS, Dr. Iezzoni pioneered research into healthcare access among people with disabilities. Recognized as a leading expert in disability studies, Dr. Iezzoni was the first female researcher affiliated with Beth Israel

Deaconess Medical Center to be appointed as full professor at Harvard Medical School.

"For many disability communities, having digital access has been really critical," she told me via a Microsoft Teams meeting. "I've used a wheelchair since 1988—so for 34 years," she adds, "and the COVID pandemic had some teeny-weeny threads of silver lining where it's really pushed us into technologies that we knew about but hadn't actually been taken to their natural next steps."

Speaking with a trace of firmness, Dr. Iezzoni clarifies: "For people like me who cannot walk, telehealth is great because I don't have to get on the subway system, I don't have to worry about the weather or having a lap full of rainwater, which is, by the way, one of the worst things."

As a board member of a group called Boston Center for Independent Living, Dr. Iezzoni listens to lots of conversations about disability and digital care. "Someone who is deaf, for example, who wants to communicate with a hearing person, might have to go through a telecommunications relay service—where you have an interpreter reading their sign language and then voicing it to the person on the other end of the phone." The potential to do away with this via immediate captioning during video visits, she describes as "hugely consequential."

Dr. Iezzoni intimately understands that the mere presence of accessible equipment won't ensure it is used, or that patients will be treated fairly. She wrote a book about it—*More than Ramps: A Guide to Improving Health Care Quality and Access for People with Disabilities*—describing the variety of structural and discriminatory barriers arising in healthcare for people with disabilities, many of which also affect seniors.[33]

After listening to patients' and families' views on how telemedicine reduces mobility challenges, she says that allowing patients to stay in their own homes can help those with intellectual disabilities, autism, or communication difficulties avoid numerous hidden barriers:

"In-person appointments for this population can be very, *very* complicated. You might need a low-light environment or an environment with low noise. You might need to prevent people from having to sit in a waiting room, you might have to put them into a quiet space before they can come in with a clinician." Dr. Iezzoni cautions that for some people, including those with serious intellectual or cognitive disabilities, problems like these can cause genuine confusion or anxiety.

Yet patients who are speech-impaired or blind, or those with manual dexterity problems, still struggle with video-based technologies. At the time of writing, many telemedicine platforms do not yet operate with custom features, and patients must now await all-important "accommodations"—this time virtually. Nor does telemedicine do away with the need for lab tests or other on-site necessities which are often conducted immediately after a traditional, in-person appointment.

New freeways to physicians

"It's heaven knowing I don't have to travel for two and a half hours for a 30-minute office visit. It no longer takes literally an entire day and then another day of recovery for a half hour visit. . . . The other big thing," says Jen, "is that it's given me access to doctors I wouldn't have been able to see as easily or as often."

Since adopting telemedicine during COVID-19, more than 80 percent of American doctors say patients have better access to care.[34] Reductions in travel time and financial savings are the most frequently cited advantages of telemedicine. In 2017 a California-based study of 19,000 appointments with 11,000 patients found telemedicine spared patients a total of 5 million miles and a whopping 4.7 million minutes in travel time—the equivalent of nearly nine years spent getting to and from doctors' offices.[35] Patients saved an average of $156 per appointment. Conducted between 2018 to 2019, a study investigating data from 2.2 million primary care office visits to the Kaiser Permanente health system in Northern California found patients were much more likely to opt for telemedicine—whether telephone or video—if their clinic required them to pay for parking fees.[36]

During the later stages of the pandemic, a study led by Rebecca E. Anastos-Wallen and Krisda Chaiyachati at the University of Pennsylvania's Perelman School of Medicine found a narrowing of racial disparities in appointment attendance: in March 2020, Black patients completed 60–63 percent of primary care appointments but by the end of 2020 this rose to 68 percent.[37] Even as early as 2003 a review of 306 studies concluded that remote visits offered "significant socio-economic benefit to patients and families," including "increased access," and "better quality of life."[38] Since then, multiple findings demonstrate that telemedicine can help bulldoze barriers for people with lower incomes, those holding down multiple jobs, parents juggling multiple duties, and patients in rural areas.[39]

Given telemedicine's undoubted potential to expand access to doctors, why did healthcare not harness it sooner? While many doctors empathize with patients' struggles, they often fail to see accessibility as their responsibility. Worse still, doctors sometimes perceive missed appointments as a moral failing on the part of patients.

I put the question to Dr. Jorge Rodriguez, an internist at Brigham and Women's Hospital in Boston, Massachusetts. Dr. Rodriguez is an expert on telemedicine and healthcare equity; he was diplomatic but blunt: "On the healthcare system side, even though we knew that some patients were struggling to get to visits, or parking was very expensive, or it was impacting their ability to work, it wasn't enough of a pain point to say, *'Okay, we need to move to telehealth to make it easier for people to access care.'* It wasn't enough of a pain point to move fast."

Equally, he told me, patients were largely in the dark about what telemedicine could do for them: "You go to your doctor's office, and you sit in the waiting room. On the patient side, it was, *'Okay, this is just the way things have always been done.'*"

Dr. Lisa Iezzoni echoes this view: "I remember ten years ago people said, '*Oh, we won't have to travel anymore, we'll just teleconference.*' But it never happened—not until the pandemic."

For Jen Lawson, the pain points of traditional care—costs, travel, logistics, or discomfort—were both literal and acutely felt, even if the telemedicine remedy wasn't clear to her prior to the pandemic tipping point. On the medical side, many pieces of the puzzle were not yet in place: hardware, devices, electronic infrastructure, and a finer-grained awareness of telemedicine's clinical value—which we'll turn to shortly.

From the perspective of overworked doctors, deep fears about change—including "technostress"—are also understandable. Disruptions to established ways of working can involve serious upheavals, especially in already pressurized environments. In her study, Donna Zulman conducted nearly 70 interviews with specially appointed VA health staff responsible for implementing telemedicine.[40] She identified staffing shortages and inadequate training as key barriers to adoption.

On the other hand, change is additionally hard when doctors and other actors have financial skin in the game. Although not the case in many health systems, this featured as an additional concern in some practices and countries. Since virtual visits are cheaper and faster than traditional appointments in private practices, telemedicine potentially reduces doctors' billing and reimbursement fees. In the US, telemedicine was held back by insurance companies reluctant to pay for remote consultations, and interlocking vested interests among key players unquestionably retarded telemedicine's growth. Since the late 1800s, every American state has commanded authority to license its own doctors, an arrangement that is highly lucrative for all involved stakeholders, except the patient.

State-specific licensing provides market power to in-state physicians, who sit on medical boards and comprise a powerful lobby in state governments. For example, in 2019, California received

$62 million in doctors' applications and renewal fees. In the US, federal emergency legislation enacted during COVID-19 relaxed regulation that thwarted telemedicine but it expired on May 11, 2023. Significant strides were made in expanding telemedicine during the COVID-19 pandemic. At the time of writing, challenges around licensure and reimbursement persist—though there are some hints this might change.

Pre-pandemic, telemedicine was not routine even though its potential was well recognized. There are no silver-bullet explanations for delayed roll-out but when the history books are written, conservatism must be part of the picture. Apathy toward change was universal. For example, in 2008, an EU Commission communication entitled "Telemedicine for the benefit of patients, healthcare systems and society" noted, "Member States have long realized the potential of telemedicine and are supportive of its beneficial deployment. However, despite this support and the considerable level of technical maturity of different technologies, the sector is not as well developed as could be expected."[41]

In Britain, an NHS report published in 2011 found that "active and passive" resistance by doctors "stifled" telemedicine's roll-out, yet failure to implement it, the report warned, would result in "disastrous consequences" for public health.[42] By December 2019 a survey by the Royal College of Physicians found that most British doctors had still not conducted any video visits with patients, despite 70 percent admitting that at least some appointments could be done virtually.[43] More embarrassing, a quarter of those surveyed admitted they'd never telephoned a patient, with around half saying that fewer than one in ten consultations were conducted this way. Pre-COVID, some British doctors hadn't entered the era of Alexander Graham Bell, never mind the epoch of the internet.

Meanwhile, in 2019, a US survey found that despite patient enthusiasm for telemedicine, a key reason for lack of uptake, cited by 40 percent of respondents, was "their physician does not offer it."[44] In

Donna Zulman's study at the VA, provider disinterest was the third leading barrier to implementation. On this point, Dr. Rodriguez is forthright: "We just didn't have enough of an impetus to change until we had to. Aside from money and payment drives, which affect a lot of the decisions we make in healthcare, the other major driver was just general comfort." The comfort was for the providers.

Digital ditches

Jen Lawson is digitally savvy. She also owns an iPad, and lives in a home with high-speed internet. Not everyone is as fortunate. A variety of digital ditches can derail patients' access to telemedicine. The affordability of smartphones, laptops, and stable home internet access determines who gets to reap the benefits of convenient video visits. So, although it reduces personal expenditure and travel time, it would be wrong to say this recently opened freeway to physicians is frictionless—or indeed free.

The good news is that the divide between those who can meaningfully engage with internet-based technologies and those who cannot is closing. Worldwide, year on year, access to the internet and digital devices is curving upwards. Globally, in 2019 almost 70 percent of youth—aged 15 to 24—used the internet.[45] By the end of 2023, 4.6 billion people—57 percent of the global population—used mobile internet.[46] The bad news is that, for some, there is still a veritable tech chasm to vault. People living in remote or rural areas often experience poorer internet connectivity. In 2024, 2.5 billion people worldwide lacked internet access.[47] In low- and middle-income countries, rural adults are 28 percent less likely than urban residents to access mobile internet, and women are 15 percent less likely than men to use it. Despite mobile broadband availability, 39 percent of people don't use it, showing that access alone doesn't guarantee adoption.[46]

In the US, Black and Latinx households are less likely than white homes to report access to high-speed internet.[48] Digital divides also exist among the elderly, people living in rural areas, those with low incomes, and individuals with fewer years of formal education, who are also less likely to own a smartphone, or use the internet. However, within countries, the deepest digital disparities may be associated with disabilities. Significantly fewer people with disabilities own digital devices, can access broadband, or report using the internet.[49,50]

There are consequences to this divide. Lack of "digital capital" is recognized to be a determinant of health.[51] In the US, for example, multiple surveys—conducted both before and at the beginning of the pandemic—revealed older people, patients who were poorer, non-white patients, those living in rural areas or areas without broadband, and non-English speakers were less likely to use video-based telemedicine.[52,53] The medium can open doors or create barriers for older people, largely depending on patients' comfort with digital tools and the level of social support they receive. Access hinges on both tech skills and having supportive networks who encourage use.[54] Among older people, auditory and visual impairments can compound the access challenges.[55]

Closing digital gaps would expand healthcare access, ensuring more people get the care they need. In the wake of the 2021 Infrastructure Investment and Jobs Act, Dr. Rodriguez was cautiously optimistic for Americans.[56] While not focused expressly on healthcare, the aim of the act was to reduce digital disparities by earmarking $65 billion for digital inclusion initiatives. In an interview with *The New England Journal of Medicine* in 2022, he broke it down: "There's about $42.5 billion, which specifically invests in actual broadband infrastructure. So, the laying down of wire or satellite, whatever it is, to actually build out internet access in communities that may lack it."[57]

Throwing dollars after digital devices and internet access will be futile, however, if patients don't know how to get online. Surveys in the US and Europe show that older adults are willing to adopt technologies but need advice on how to use them.[54,58] The Infrastructure Investment and Jobs Act specifically allocated $2.8 billion to build digital literacy programs, aiming to help more patients access health portals and take advantage of video visits with doctors.

Similarly, launched in 2021, the EU's "Digital Decade" program is a series of policies and targets aimed at improving internet infrastructure and connectivity, and digital literacy among its citizens by 2030.[59] These policies have made a tangible impact but remain fragile and can be easily undone by administrative shifts. In addition, they cannot address barriers such as clinician biases. Research demonstrates, for example, that some people are excluded from video visits even when they have the means to access doctors this way. A study led by Ivy Benjenk at the University of Maryland School of Public Health found that, during the pandemic, 35 percent of Medicare patients were only offered telephone-based telemedicine.[60] Yet two thirds of these patients owned smartphones or tablets, and could access the internet at home, meaning they might have availed themselves of video visits if invited. Excluded patients were more likely to be Black, Hispanic, or non-English speakers, or to live in rural areas.

Research led by Dr. Rodriguez echoes the demographic exclusions found by Dr. Benjenk and her team. In a study of nearly 1,700 primary and specialist care practices at a major US healthcare system, he found that 38 percent of the variation in video visits was associated with the clinics at which patients received care, and 26 percent with clinicians' decisions about who was to be offered video visits.[52]

"So, even if you had a patient that was fully connected," he explained to me, "there was something about the clinic and the clinician delivering the care that swayed whether they're going to do video versus audio, or even telemedicine versus face-to-face visits."

"Perhaps clinicians were approaching patients and making some pre-judgment, assuming '*Well, you probably are not going to use the video visit. So, I'm just going to do audio with you.*'"

Overcoming discriminatory biases among doctors is something we'll drill into later in the book. For now, the successful expansion of telemedicine requires electronic health vendors, digital designers, and healthcare systems, including state-run organizations, to collaborate with diverse patient populations. Working alongside the elderly and people living with disabilities these actors could better ensure the technology is adequately tailored to serve a wider variety of needs.

Dr. Rodriguez tells me his biggest concern is health organizations hyping themselves as "digital first" without putting in the necessary work: "I'm afraid inclusivity will get lost, and we'll start recreating that divide—that we'll go back to '*Oh, this is just the way we deliver care.*'"

New hazards

David Nash was a law student living in Leeds, in the north of England. During the pandemic, on October 14, 2020, he called his primary care practice over concerns about lumps on his neck. On October 23, David reached out again, this time about a painful, hot right ear. A few days later, on October 28, he spoke to a GP about blood in his urine and was diagnosed over the phone with a urinary tract infection. Following yet another telephone call on November 2, David was diagnosed with a flu-like viral infection.

Over a 19-day period, he spoke with his GP practice four times. Two days after his fourth phone call, David died from brainstem swelling at Leeds General Infirmary. In January 2023, following an inquiry, a coroner ruled that David Nash suffered from mastoiditis, a serious bacterial infection. Medical errors arise during in-person appointments too, but in this case the coroner's conclusion was the 26-year-old student would "likely have lived" if he had been offered the chance to see his GP in person.[61]

David Nash's tragic case underscores the findings of a UK study, which revealed that while such remote safety incidents are rare, they can still result in severe harm or even death.[62] Post-pandemic, many medical organizations—including the World Health Organization, the American Medical Association, and the NHS—issued guidance about the use and implementation of telemedicine. This guidance is aimed at improving patient safety, in deciding when telephone or video visits, or in-person care, is superior. A major worry is that video and phone appointments constitute second-class, or even third-class care, respectively.

Dr. Jorge Rodriguez offered me his opinion: "Video is generally preferred to audio. As doctors, we want more data and video gives you another data stream. But the truth is, there are still more questions than answers."

Qualifying his response, he continues: "Sometimes, you need something really intense for a job and sometimes something simpler will do. If the main goal is, say, I want to chat with you about your diabetes control, telephone can suffice. If I'm just asking: *'What were your readings?'* or *'We'll make these adjustments—sound good?'* telephone is fine and a case of, *'Great, off you go.'*"

Face-to-face meetings allow doctors to obtain a fuller picture: "I work as a hospitalist and on the inpatient side, one of my big things as I approach them is to give patients that first look. I can get a very quick sense: does this patient look unwell, and how concerned am I about them? Certainly, when you're relying purely on a patient's voice you miss that."

The pandemic provided greater evidence, at scale, of the strengths and weaknesses of telemedicine. A study published in 2022 found that video-based telemedicine use was associated with either significantly better performance or no difference in 13 of 16 measures of primary care quality.[63]

In other research, conducted at the Mayo Clinic, researchers compared video with face-to-face appointments. During the early stage of COVID-19, the Mayo team investigated the diagnoses of nearly 2,400 patients who underwent a video visit which was later followed up by an in-person meeting. Their findings are reassuring but underline the need for caution.[64] Across all branches of medicine, they found that 87 percent of video diagnoses were later supported by doctors' judgments in face-to-face encounters, but accuracy rates varied according to specialty. Confirming the well-documented benefits of telemedicine in mental healthcare, diagnoses made during psychiatric video visits matched traditional in-person care an impressive 96 percent of the time. However, video visits in primary care matched in-person appointments only 81 percent of the time: meaning nearly one in five diagnoses were later overturned by a doctor seeing a patient face to face.

The Mayo team also found that, among new patients or those requiring more hands-on medical examinations, rates of diagnostic accuracy decreased. Patient age, and potentially disability, also stood out: the odds of accurate diagnoses in video visits fell by 9 percent for every ten-year increase in patient age. According to the investigators, dips in accuracy might reflect older patients experiencing greater difficulties with the technology or enduring higher rates of vision and hearing problems when they do get online.

What does telemedicine mean for medicine's upside-down access problems? In October 2022, Dr. Adrienne Boissy, Chief Medical Officer at Qualtrics, and former Cleveland Clinic Chief Experience Officer, gave an unequivocal answer to *Healthcare IT News*: "With PHE [public health emergency], we saw the industry put patients and their access to care first—no longer hindered by location or demographics.... To revert back to reimbursement models that only support in-person care unravels the gains of meeting people where they are—physically and emotionally."[65]

In 2022 the World Health Organization issued a news release on an investigation of findings drawing on more than 20,000 studies across 53 member countries.[66] The study, which enrolled more than 20,000 patients, concluded that there is "a clear benefit of telemedicine technologies in the screening, diagnosis, management, treatment, and long-term follow-up of a series of chronic diseases."[67]

The medium is not an access panacea. It doesn't magically eliminate all barriers to traditional care. In relative terms—comparing telemedicine with in-person appointments—there are also no guarantees it can fully right medicine's disparities problems. Recall the study by Anastos-Wallen and Chaiyachati at the University of Pennsylvania, which found telemedicine increased appointment attendance among Black patients largely because of an uptick in telephone appointments—which can be an inferior method of clinical consultation.

Skeptics might wonder whether, on balance, healthcare's long-held apathy toward telemedicine was therefore justified. Due diligence in addressing the novel problems presented by telemedicine is crucial. But dismissing the medium's considerable benefits would miss the bigger picture and neglect the most marginalized patients. For many would-be patients, the correct comparison is not telemedicine versus in-person care, it is telemedicine versus no care.

During our Google Meet call, slowly turning to plump the pillow behind her Jen Lawson told me, "It's pretty simple. Telemedicine makes my life easier and less stressful." Compared to a two-and-a-half-hour round trip which took a day to recover from, that, at least, is a start.

PART II
DISCLOSING OUR SYMPTOMS

4

Doctor Deference

When he walks into a room, Professor Ernest Thomas Lawson—a youthful 93-year-old—is used to people looking up. Like his daughter Jen, he is gentleness embodied in a towering physique. Nowadays his 6' 4", angular frame is slightly bent from stooping—all the better to chat with his numerous friends, many of whom are former students.

Tom—as he is known to friends and family—has lived in Kalamazoo for 70 years but still speaks with a distinctive South African accent. Born in Cape Town to a poor white family, and the youngest of nine children, Tom was brought up on the edge of the Kalahari Desert. At the age of 18, "with a headful of dreams," he bought a one-way ticket to America. "Of course, I soon realized 'the land of the free' wasn't all Coca-Cola and milkshakes," he tells me, "but man was it exciting!" Ten years after arriving he graduated from the University of Chicago with a PhD in philosophy and secured a tenure-track position at Western Michigan University. He confides that his "real pride and joy" was purchasing a Jaguar E-Type. A poster-boy for the American dream, Tom describes himself as "an incurable optimist but no fantasist." As the highly esteemed "Father of the Cognitive Science of Religion," Tom Lawson knows better than most the idiosyncrasies of the human mind.

Today he is wearing a bright orange shirt, slim black jeans, and his signature tan cowboy belt. He conveys all the vitality and enthusiasm of a man whose biggest life adventures are still ahead. Tom tells me his deepest regret is turning down lecture invites. Ten years ago, this was unavoidable.

"Three hours before my flight to Europe, a searing pain literally stopped me in my tracks. My knee gave up. I never experienced anything like it—the airport staff had to wheel me out of duty-free and I wound up missing my trip."

The very next day Tom's family doctor referred him to an orthopedist—a specialist in joint and bone surgery—and a few days after that, on a scorching September morning, Tom met Dr. Parker.

"Naturally, I was nervous, but this guy was just terrific. Immediately I knew I was in good hands." Tom continues, "Parker was a top guy. Of course, he was in demand, so I was happy to wait. The waiting room was very ritzy with freshly brewed coffee and everything."

"*And man, what an office!* He was sitting behind this beautiful oak desk. Anyway," Tom interrupts himself, "we talked about my osteoarthritis. There was no question, Parker said, I needed a knee replacement. We discussed every step of the procedure. Next time we met, he was wearing his scrubs."

After the surgery, the nurses informed Tom the procedure went well. Quietly, however, he felt otherwise. Lowering his voice slightly, Tom says: "The trouble was, soon after waking, I noticed I couldn't pee. Or I could—sort of—but it took a while and even then, my bladder wouldn't empty."

Scrunching his nose, he told me, "You know, it was a bit *embarrassing*."

Eventually, Tom plucked up the courage to tell a hospital nurse who advised him the problem was quite common after sedation. On his busy rounds, Dr. Parker also caught up with Tom: "Of course, his chief concern was the knee, and to ensure everything had gone to plan. But he confirmed anesthesia can cause urinary retention and the symptoms would soon go away."

Back home, Tom worked tenaciously on a recommended program of exercise to strengthen his knee. "At the next appointment I was determined to show my progress. But, in truth," he admits, "I still had the peeing problem, and it was getting worse. Parker is a busy guy, so I didn't like to hassle him before my next check-up."

Two weeks later, delay was no longer an option. Tom finally called his nurse practitioner. "She was very disappointed—*Why had I not phoned her sooner?* You know, she was right to scold me."

Tom's prostate was so swollen it had squeezed the urethra shut, causing urinary blockage. Removing nearly 3 liters of urine, his doctor told him delaying by even half a day more could have risked kidney damage. Diagnosed as "BHP"—benign prostatic hyperplasia, a common condition among older men—his symptoms were likely unrelated to his knee surgery. But one fall with his fragile knee and his bladder could have ruptured and killed him.

Tom Lawson's health scare sounds like a series of wacky flukes that doctors would need a crystal ball to avert. However, his case is both more predictable and emblematic of medical appointments than it first appears. In this chapter we'll discover how Tom's story exemplifies a common social syntax in medicine. I call this cluster of problems "doctor deference."

Dying of embarrassment

After accessing healthcare, the next stage of our medical journey is fully disclosing our problems to health professionals. Like Tom, many of us are medical procrastinators and there are multiple reasons why. Lacking clinical expertise, sometimes we aren't clear when symptoms are trifling niggles or something serious. Tom knew something was off, but he didn't know when it was prudent to seek help.

Another reason for Tom's delay was saving face. The desire to keep up appearances is very human and normal. In 1956 Canadian sociologist Erving Goffman coined the phrase "impression management" to describe the tactical behavior involved in presenting a positive image of oneself in the company of others,[1] and today psychologists refer to this as "social desirability bias."

Fifty years after Goffman described the phenomenon, a study in the US investigated whether patients employed such social tactics in health settings.[2] Participants were asked: "In order to get the best treatment possible at the doctor's office, how important do you think it is for you, yourself, to wear nice-looking clothes to an appointment?" Other questions probed the importance of looking clean, arriving on time, being friendly with the doctor, letting the doctor know you care about your health, and showing you are an intelligent person. Results overwhelmingly showed that people considered positive self-presentation to be very important for receiving the best medical care possible. In keeping with medicine's inverted traditions, patients who were poor, Black, female, or elderly were even more likely to believe positive presentation was valuable.

Monitoring one's appearance and demeanor compromises concentration.[3] Focusing attention on how they are perceived, patients may be more likely to forget what questions to ask, and pretend they understood what the doctor said.

Preserving one's image also inclines patients to massage the truth. Hippocrates wisely advised: "keep a watch also on the faults of the patients, which often make them lie about the taking of things prescribed."[4] More recently philosopher of medicine Miriam Solomon summed it up this way: "patients misremember, fabricate, exaggerate, lie, distort, selectively tell, or otherwise intentionally or unintentionally, explicitly or implicitly, report falsehoods," often "in an attempt to present themselves in a favorable light."[5]

Patients are doctor-pleasing purveyors of evasion, avoidance, and even misinformation. Many of us can't resist telling the odd white lie or the occasional whopper.[6,7] In his diarized memoir *Trust Me: I'm a (Junior) Doctor*, Max Pemberton recounts one such fantastical fib: "I don't know what is more ludicrous," he wrote, "that someone has just come into A&E with a hairbrush inserted into his rectum or that he is expecting me to believe that he 'fell' whilst brushing his hair and as he put his hand down to break his fall, it lodged in there that way."[8]

Subtler finagling of the facts is more common. One US study found 85 percent of patients admitted to concealing or massaging the truth, and around a third outright lied to their doctors.[9] Another survey of 1,200 American patients found nearly half "sometimes" or "often" lied to their doctor.[10] Top fake news related to following physicians' advice, diet and exercise habits, booze consumption, and medically relevant information about their sex lives. One in two admitted embarrassment was the reason for "alternative facts."

Exploring the problem further, recently psychologist Dr. Andrea Levy conducted studies which revealed patient self-censorship is common. Across two online surveys of more than 5,000 participants, 61 to 81 percent admitted withholding medically relevant information.[11] Like Tom, confusion about how to interpret health issues was a common reason for nondisclosure, with around a third staying schtum because they "didn't think it mattered."

Tom was anxious and embarrassed. In primary care, some patients (my father was one of them) spend their visit discussing the bothersome ingrown toenail only to mention the blood in their stools as they leave the office. Similarly, Levy found that between 50 to 61 percent of respondents confessed that "embarrassment" lay at the root of their reluctance to speak up. Overall, the top reason for hiding information, among 64 to 82 percent of surveyed participants, was "I didn't want to be judged or get a lecture." Around four in ten stayed silent because they "didn't want the healthcare provider to think I'm a difficult patient" or because they "didn't want to take up any more of the healthcare providers' time," with about three in ten reporting "I didn't want the healthcare provider to think I'm stupid." Finally, demonstrating a kind of lickspittle logic, 16 to 21 percent said they withheld because they "wanted the healthcare provider to like me."

Fear of embarrassment in front of physicians is easy to make light of, but it can be lethal. An international review published in *The Lancet* probed the reasons why people delayed seeking medical attention for at least twenty different kinds of cancer.[12] Analyzing 32 papers published between 1985 to 2004, investigators identified recurrent, cross-cultural reasons why people experiencing serious red flag symptoms procrastinated.

Like Tom, a leading reason for stalling was lack of health literacy: some reported being so anxious they couldn't bear hearing their worst fears confirmed. Strikingly, however, whether they lived in Holland, Hong Kong, or Hawaii, would-be patients were remarkably self-conscious: "men and women worried about being labelled as neurotic, a hypochondriac, or a time-waster."[12] Chronic fear about vexing doctors, and sensitivities relating to symptoms and body parts, posed persistent barriers. Some sought to legitimize their help-seeking by waiting for another reason to make an appointment. For some the fear was overwhelming: "I was lucky," one patient recollected, "I didn't have to go to my GP because I collapsed in church."

Another UK study drew similar conclusions, finding that many people experiencing "cancer alarm symptoms" were preoccupied with being viewed as a "nuisance."[13] Ominously, in a three-month follow-up study, pathological self-consciousness remained a key deterrent; one interviewee attested: "[I'm] not scared of what he [GP] might think of me, exactly, but I don't want to bother him."[14]

Some worries about encroaching on doctors' time are driven by the pressures on health systems and public purse strings.[15] But this doesn't explain all of patients' diffidence. A multinational survey of nearly 20,000 participants in Australia, Canada, Denmark, Norway, Sweden, and the UK found that, wherever people lived, at least 10 to 15 percent said they would be too embarrassed to ask their doctor to check serious symptoms, including lumps or swellings, unexplained bleeding, or changes in bowel or bladder habits.[16]

It's not only cancer care that is affected. Other socially sensitive and stigmatized conditions are prone to patients' red-faced reticence. A US study of more than a thousand patients with major depression found that 43 percent offered at least one reason for delaying help-seeking,[17] with patients worrying "I might cry or become too emotional during the visit," "The doctor might think less of me," or "I would not know how to bring up the topic of depression." Andrea Levy found that around four in ten people experiencing depression, suicidality, and abuse, including sexual abuse, withheld highly sensitive information from clinicians: the chief reason was embarrassment.[18]

Explaining doctors' power

To uncover the roots of "doctor deference" and why it arises, we need to understand the origins of physicians' status. Sociologists often describe doctors as grasping for power over patients.[19] Seldom explained, however, is the nature of this power, where it comes from, and why doctors want to wield it over patients in the first place.

To obtain deeper insight, the work of psychologists Joey Cheng of York University and Joseph Henrich of Harvard University is helpful. They are leading experts on the psychology of status, including why it evolved, and how modern settings can evoke our ancient responses. Cheng and Henrich invite us to consider the management styles of two highly successful yet markedly different American businessmen: Warren Buffett, retired CEO of Berkshire Hathaway, and one of the world's richest people, and Henry Ford II, grandson of the founder of the Ford Motor Company.[20] Warren Buffett, aka "The Sage and Oracle of Omaha" is, as the epithet suggests, one of the most respected figures in the US and many make trips to his headquarters in Nebraska to seek his counsel. Buffett's reputation is for "subtly steering rather than controlling every decision-making process," inspiring trust in those who work for him.

Contrast Warren Buffett with Henry Ford's grandson. For nearly forty years, in a turbulent post-World War II economy, Henry Ford II built his grandfather's company into the second-largest industrial corporation worldwide. Ford's success, Cheng and Henrich argue, was in part attributable to his "reputation for erratic outbursts of temper," and "unleashing humiliation and punishment at will upon his employees," who viewed him as "a terrorizing dictator, bigot, and hypocrite." Notoriously, when challenged by subordinates, Ford II reminded those who dared to question him, "*My* name is on the building." Creating a culture of threat and intimidation helped Ford II establish ironclad authority.

Professors Cheng and Henrich argue that there are two ways to the top. If Warren Buffett is a poster-boy for prestige, Henry Ford II is a pin-up for dominance. Social influence can be garnered by respect and admiration or harnessed by bullying tactics and intimidation. Unfortunately for patients, doctors command both forms of authority. Before we explore the ways in which physicians behave like Warren Buffett *and* Henry Ford II—and how this can influence our willingness to visit the doctor, what we disclose, and how the visit unfolds—a brief history of status and subordination is in order.

Our ancestors lived in small bands of hunter-gatherers where there was little doubt about who had useful skills—for example, who was the best hunter, who excelled at sourcing edible plants, or who was the wisest diplomat.[21] Natural selection equipped humans to pick up on life-saving competencies by inclining them to cozy up to these targets. According to evolutionary psychologists, mimicry is the sincerest form of flattery, and a "prestige bias"—a capacity for useful imitation—enabled our ancestors to identify proficient individuals, sidle up to them, and ingratiate themselves into their company—all the better to obtain a front-row seat to copy, learn, and master their prized skills.[22]

In their work, Cheng and Henrich identify key patterns of behavior associated with both the prestigious target of copying (the "higher-up"), and those who mimic them (the "lower-status" party). Low-status parties tend to show admiration, awe, and genuine respect for the prestigious individual, automatically engaging in direct gazing, attentive listening, and nonconscious imitation. Their behavior is also marked by linguistic signatures such as giving the floor to the higher-up, permitting long conversational pauses, and expressing agreement.

Doctors are prestigious people. It is precisely because of their life-saving skills and knowledge that we confer this status on them: throughout the world, doctors routinely top polls of the most respected professionals. By default, then, prestige is built into traditional medical appointments. This observation may seem banal but, from the standpoint of the psychology of status, it carries consequences for our behavior. While we may balk at the comparison, or even deny it, in our medical appointments, doctors' contextually higher status tends reliably and predictably to turn us into admiring and respectful lower-status subordinates. When it comes to seeking medical advice, the power of prestige can be positive: when we respect our doctor, we take what they say seriously. Tom embarked on his knee exercises, keen to impress Dr. Parker at his next appointment.

However, medicine exhibits a blind spot when it comes to the negative aspects of doctors' prestige. Too much respect and adulation can be a bad thing, and, through no fault of their own, doctors' vaunted status can incur nosediving downsides for patient dialogue. Traditional medical appointments are a social setting that can inadvertently gag the honesty of patients: doctors' prestige can meddle with disclosures influencing what, how, when, or even whether we divulge information.

The psychology of medical prestige manifests in other ways. When not awaiting scary news, or if symptoms are not life-threatening, patients may feel a bit of a buzz sitting down for a one-on-one with their doctor. So strong is the desire for proximity and good favor, lower-status parties are inclined to offer tokens of appreciation. Doctors are often the recipients of generous gifts. Such is the extent of patient gift-giving that medical organizations implement strict rules limiting what doctors can reasonably accept. In the UK, the British Medical Association stipulates that doctors should keep a register of gifts worth £100 or more from patients or their relatives. The American Medical Association requests doctors to "decline gifts that are disproportionately or inappropriately large."[23] Rules like these don't pop up in low-prestige jobs.

In our ancestral habitat, the mere trappings of success provided indirect but reliable clues about a person's acumen. Access to surplus food, scarce resources, and other costly or unique possessions stood as Stone Age billboards of superior skill and ability, and therefore of status. Fast-forward to the modern medical environment, and material garb and goods influence our impressions of doctors' rank and standing. Casting his eyes over Dr. Parker's waiting room and office, Tom was awed by the environs in which he found himself, which likely ramped up his positive impressions of Dr. Parker's professionalism.

Although signals of prestige were generally reliable in the hunter-gatherer environment in which these biases were selected for, nowadays mismatches between our modern environment and our old psychology demonstrate just how easily our Stone Age minds can short-circuit. Like furtive curtain twitchers in suburban cul-de-sacs, humans are apt to keep a beady eye on what the Joneses are up to, and to be subtly, surreptitiously swayed by it. Cues of superficial material goods and fine things can influence perceptions, and research shows people have an unwitting proclivity to stereotype wealthy people, and those with expensive things, as more competent.[24]

In one study, participants were requested to rate the quality of care at different medical facilities based on 35 pictures of actual waiting rooms varying from shambolic to chic. Waiting rooms with classier furniture, brighter lighting, and more tasteful artwork enhanced judgments about clinicians' competence.[25] While medical curricula don't traditionally encompass feng shui, physicians might do well to strategically flaunt their hard-won credentials. Far from judging them vainglorious, people perceive clinicians with multiple diplomas as more competent than their humbler peers.[26]

Or consider a survey conducted by the British Medical Association in 2012, which reported that 75 percent of doctors believed they should be "free to express their individuality through their appearance."[27] Binning their work clobber, however, might diminish patient perceptions about physician competence. In 2015, a systematic review found patients preferred primary care doctors in formal attire, associating professional garb with superior acumen.[28] In 2022 the largest survey of patients' fashion preferences ever conducted across three continents concluded that patients preferred doctors in white coats above any other attire.[29]

Cheng and Henrich observed that there are two ways to the top. Alongside the prestigious Warren Buffett is the dominant Henry Ford II. Dominance has distinct origins, and psychologists propose that aggression was a winning strategy in the competition for mates and resources. Signals of submission were also adaptive. When confronted by an aggressor, it was advantageous to make a quick-fire judgment call: either square up to them, lock eye contact, and prepare for combat, *or* respond timidly, avoid their gaze, and signal defeat. When it comes to confrontation, humans have an instinctive capacity to weigh up the costs and benefits of competing. Submissiveness, it is proposed, evolved as a de-escalating strategy affording shrewd, weaker parties' survival; valiant but instinctively inferior tacticians were more likely to die before reproducing. Faced with a formidable competitor, it was strategic to hang low, bide one's time, and yield to the stronger, meaner contender.

The dynamics of dominance and submissiveness are found in all kinds of modern settings, even though social structures and cultural customs are nowadays aimed at keeping bullies in check. Dominance based on coercion and threat is typified by expansive and open body language: legs astride, hands on hips (think commanding CEO using his desk as a footstool). Dominant individuals are more likely to seize the floor in conversation, and to adopt aggressive verbal intimidation such as disparaging humor or criticism. Cheng and Henrich also report that dominance is typified by hubris, egocentricity, self-aggrandizement, and arrogance.[30,31]

In stark contrast, lower-status individuals send unambiguous signals of submission by adopting a hunched posture, a lowered head, avoidance of eye contact, and a compliant demeanor—keenly alert for their aggressor's next move.[32] Subjugated persons also tend to avoid "higher-ups," feeling trepidation or shame.

Prestigious heroes and bullying villains to one side, psychologists emphasize that human cognition and behavior is not neatly categorized: an individual might be perceived as prestigious in one context and undistinguished in another, dominant in some scenarios, meek in others. The psychology of status gets even messier. Contrary to what sociologists and cultural commentators often imply, power can be an *outcome* of prestige rather than a motivation for it.[20] People who are highly skilled—that is, considered prestigious in a particular context—may encounter more opportunities to behave badly, that is, become dominant. Put another way, prestige has real potential to turn petulant.

Even esteemed CEO Warren Buffett has greater odds of transmogrifying into a Class-A bully—a tendency, by all accounts, he seems to have resisted. However, precisely because of his success Buffett commands employee rewards and punishment. Attaining high rank through his revered abilities, Buffett has the capacity to inflict serious costs. The lesson, Joey Cheng says, is that "formerly prestige-based relationships may become more grounded in dominance."[33] We can observe this fluctuation in front of our very eyes when a prestigious person is challenged by a lower-status party,[21] with the psychology of prestige morphing into a face-off for dominance. Try it yourself, with a word of caution—as with the celebrated skirmish between intellectuals William F. Buckley and Gore Vidal, the insulted party may want to "sock you in the goddamn face."

Medical appointments unintentionally prompt subordination among patients. Subtle situational cues can elicit a subtle, cap-doffing doctor deference. These responses are automatic and nonconscious. Tom is a distinguished academic, but during his visit Dr. Parker turned him into a supplicant.

To explore the myriad of ways the psychology of status makes a direct impact on medical interactions, let's return to the appointment and the very moment the doctor greets us.

Doctors dominating the dialogue

When a doctor introduces herself, she will not say, "Hi, my name is Sandra, I will be looking after you." This is not Hooters. In fact, it's unlikely that she (or he) will use their first name. Instead, they will probably say something like, "Hello, my name is *Dr.* Smith." Most people prefer it this way. A survey published in the *BMJ* found only one in ten patients favored doctors dropping the medical prefix and using their first name alone (e.g., "Sandra").[34] Eight in ten felt awkward about reciprocating these formalities, preferring doctors to call *them* by their first name. Prefixes and honorifics provide another tip-off about prestige biases in the clinic.

After pleasantries are exchanged, we might suppose most consultations unfold with a broad question, such as: "What brought you here today?" Surprisingly, such opening inquiries are atypical. In 2019, medical doctor Naykky Singh Ospina of the University of Florida led a study that found only about one in three doctors directly invited patients to explain the reason for their visit.[35] Primary care doctors were more likely to pose an opening question—about 50 percent did so compared with 20 percent of specialists—but even this figure is conspicuously low. More striking, from the moment patients spoke, physicians tended to take the floor; even when patients were directly asked to share their concerns, doctors proceeded to cut patients off after an average of 11 seconds.

Singh Ospina's study is the latest in a decades-long chain of publications demonstrating that doctors tend to dominate dialogue, and patients tend to let it happen—exactly the kind of behavior psychologists would expect. Back in 1984, for example, a landmark study reported that in seven out of every ten primary care visits physicians interjected an average of 18 seconds into patients' opening statements, with one in ten patients derailed before they'd expressed their medical problem. Only 2 percent of patients managed to complete their initial remarks.[36] Again, in 1999, a study entitled "Soliciting the patient's agenda: have we improved?" concluded "not much." Primary care physicians cut patients off around the 23-second mark, with opening disclosures fully completed in only 8 percent of visits.[37]

Pushing back, some doctors concede interruptions are commonplace but disagree there is a problem to answer for. In 2017, writing in *JAMA*—the flagship journal of the American Medical Association—one physician noted: "some [interruptions] … improve the quality of care and help the patient make better use of time"; and that interruptions "decrease physician stress."[38] The idea that interruptions serve as an anti-stress remedy might signal hubris and dominance in its own right. However, it is reasonable to ask: might cutting to the chase improve efficiency? Findings suggest the opposite is true. Uninterrupted patients expend remarkably little extra time divulging their problems.[39] Moreover, when doctors interrupt, they usually end up dominating the rest of the conversation. In one study, 94 percent of interruptions concluded with physicians "taking the floor."[40] None of this is to say that doctors act like Henry Ford IIs during visits. Doctors do not put up their dukes or aggressively shout down or threaten patients. Nonetheless, the subtle signatures of prestige and dominance are apparent.

Doctors being derogatory

Many physicians are acutely aware that the medical profession has a powerplay problem. One of them is Harvard Professor of Public Health

Lucian Leape, who is widely regarded as the "Father of the Patient Safety Movement." In 2012, in landmark articles in the journal *Academic Medicine*, Leape and his colleagues called out healthcare's "culture of disrespect."[41,42] Leape was merciless in describing the medley of bad manners experienced by patients: "systematic disrespectful treatment includes being made to wait for appointments, receiving patronizing and dismissive answers to questions, not being given full and honest disclosure when things go wrong, and not receiving the information they [patients] need to make informed decisions."[42]

Thanks to doctors like Lucian Leape, greater attention has been paid to the dangers of incivility in healthcare. However, pointing the finger of blame at the "culture of disrespect" re-describes rather than elucidates the reasons behind these power imbalances. An evolutionary perspective affords deeper explanatory insights into why medicine's misconduct problems are so enmeshed with physicians' highly rated, life-saving human expertise. As we noted: obtaining unrivalled rank ("prestige") through their skills, senior doctors acquire immense capacity to dominate those who depend on them.

Disparaging humor is a common marker of dominance. In medicine, uncouth colloquialisms are well-documented.[43,44] In 1978 in his bestselling novel *The House of God*, Dr. Samuel Shem introduced the public to the kind of slang physicians used. "GOMER" stands for "get out of my emergency room," referring to patients with "complicated but uninspiring and incurable conditions,"[45] "to turf" means to find any excuse to refer a patient to a different department or team, and "to bounce" to return a "turfed" patient to his or her first department.

In his book *The Secret Language of Doctors*, Dr. Brian Goldman described 300 phrases and acronyms of covert clinical slang.[46] Goldman says this secret lingo is spoken by doctors of all ages and generations, including by his most trusted colleagues. Among the least offensive terms, "blabber" means "a patient who talks excessively," "PITA" is short for "pain in the ass," and "DIAL" stands for "dumb in any language."

The subtle, and sometimes bold, imprint of doctor dominance is also evident in medical records. Clinical records are meant to offer an objective account of the patient's health, but analyses indicate the ways in which physicians' authority is stamped on documentation. Phrases such as "I have instructed him to ...," "She was told to discontinue," and "patient refuses/denies/is noncompliant"[47] are an intrinsic feature of doctors' medical writing style. Beyond this, words and linguistic devices housed in records can convey derogatory attitudes. In 2021, researchers at Johns Hopkins University found stigmatizing linguistic features such as the use of quotation marks (e.g., "patient had a 'reaction' to the medication") and judgment words (e.g., "patient insists," "patient claims") in electronic records.[48]

As we noted earlier in the book, there is also the question about who gets access to these records and the debacle around online patient access illustrates doctor dominance. Allowing patients to read their own clinical records fosters a more equal partnership.[49] We conceded that physician fears are understandable. Yet some pushback may also derive from status-based responses and the moral affront that patients shouldn't be sticking their noses into doctors' business. This has certainly been the perception of researchers.

My colleague Dr. Tom Delbanco, a primary care physician, co-founded "OpenNotes" with former critical care nurse Jan Walker, at Beth Israel Deaconess Medical Center, Boston, giving a name to the transformative patient practice of online record access. During early, exploratory studies, the pair say they were not greeted with unbridled positivity by their medical peers. "A lot of doctors told us we could go to hell," says Dr. Delbanco, who also recalls a fellow doctor and trustee of a major unnamed US health organization declaring, "You are going to do this over my dead body."

Even in progressive Scandinavia, patients' access to their own medical records sometimes proved a democratic bridge too far. In 2012 researchers responsible for implementing a pilot project exploring the innovation were treated like would-be housebreakers, a description that is not mere hyperbole. My colleague Dr. Maria Hägglund, who has been at the forefront of online record access in Sweden, described what happened during the trial run at Uppsala University Hospital.

"It was controversial from the get-go," Maria told me. "Several attempts were made to block the project by local doctors and their trade union. The scheme was reported to no less than nine different authorities, including the Swedish Work Environment Authority."

In scenes resembling a Nordic noir crime show, even the police were called.[50,51] Confrontation also turned threatening: "People working on the project were accused of all kinds of serious misconduct—accusations ... [of which they] were later cleared." Maria adds, "We used to say, 'no Swedish project has ever been through such detailed ethical review.'"

Patients as subordinates

Across the world, medical records are largely considered to be the domain of doctors. Yet during our visits, we forget about half of what the doctor communicated; those who are older, in shock, or who worry more about the "presentation of self" remember less.[52,53,54] Embarrassed or anxious about encroaching on providers' time, and plain curious to know more, many look to the internet to fill in the gaps where health searches are reportedly second only to porn searches. Surveys show that up to eight in ten American adults, and about half of European patients, use the internet to learn more about their health.[55,56] In a 2024 survey commissioned by UserTesting, 67 percent of the Americans sampled said they had specifically looked up their symptoms using internet search engines such as Google or WebMD, and 52 percent had used large language model (LLM) chatbots like ChatGPT to do the same.[57]

Notwithstanding that every Google search seems to point to cancer, most of us—studies show—are savvy about the dangers of descending down cyberspace rabbit-holes. Studies conclude that even so-called "digital immigrants"—individuals born well before the dawn of the internet—are as mindful as doctors about the limitations of self-diagnosis.[58]

Doctors sleuth online too. One Harvard-affiliated primary care doctor confessed to me that he left the office to conduct Google searches to save face in front of his patients. Doctors are adopting other internet tools too, including LLM chatbots.[59,60] In my research in the US and UK, I discovered that doctors also fully suspected patients are using these tools to better understand their medical records.[60,61]

However, doctors disagree about whom internet searches or chatbots can help. Back in 2017, Dr. Sharon Tan of the National University of Singapore led a systematic review of the impact of patient web searches on doctor–patient relationships. Synthesizing the findings of 18 studies, Tan and her colleagues found significant advantages of online sleuthing: patients experienced a heightened sense of control, greater confidence, a sense of empowerment, and felt better prepared for appointments.[62] Benefits extended to increased participation in consultations, enhanced understanding of the illness or condition, and a sounder grasp of what their doctor said.

How did doctors react to patients surfing on their turf? According to patients, some welcomed it—a finding associated with visit satisfaction. However, many doctors were disparaging or affronted. Studies reported: "resistance from doctors to patients bringing up new information";[63] physicians discouraged patients asking questions from their internet search, giving them the impression that they were "disapproved of";[64] and "some health practitioners sought to assert their authority by dismissing the patient's acquired knowledge."[65]

No doctors howled, "I'm the one who does the thinking around here." Still, many patients, research shows, instinctively enacted dutifully submissive roles, preferring to humor health professionals. Rather than admitting to a little homework, some remained resolutely tight-lipped about web-sleuthing, "mindful in ensuring that doctors played the leading role during consultations."[62] Patients also worried doctors might feel insulted, criticized, or negated; some worried that their searches "might offend" or "were hesitant to ask questions" in case doing so "might displease the doctor."[66,67]

Other stealthy strategies included asking questions without explicitly disclosing details of internet searches, or silently verifying the doctor's advice against their own online findings. In a self-protective ploy, some patients admitted they would only discuss what they'd read when accompanied by family members. Perhaps, faced with physician dominance, patients revert to the perception that there is safety in numbers.

Fearing physicians

In her memoir *Dear Life: A Doctor's Story of Love and Loss*, British doctor Rachel Clarke candidly admitted, "[A]s a patient, I am meekness personified. Turn this way, look up to the ceiling, straighten your arm, open your legs, just pop up on this couch. Patients oblige, comply, obey; they cannot risk dissent when so much power is concentrated in medical hands."[68] When veering too far from acceptable roles, many of us worry doctors will be unforgiving.

In a study published in *Health Affairs*—a journal described by *The Washington Post* as "the bible of health policy"—Dr. David Frosch and his colleagues explored patients' perspectives on shared decision-making during medical visits.[69] Frosch and his team were particularly struck by a fear that questioning doctors' recommendations could lead to penalties. Recruited from the wealthy suburb of Palo Alto, Northern California, participants were affluent and well educated, yet "felt they were vulnerable and dependent on the goodwill of their physicians."

One interviewee summarized the problem: "part of the issue with doctors being gods is if you disagree with him, you're challenging [a] god … and you don't want to do that 'cause it's not a very safe thing to do." "It's difficult to fight a doctor," reported another: "You're worried … you're going to piss the doctor off … [that] it's going to change the relationship.... I don't want to rock the boat." Or, as one interviewee queried, "Is it going to come out in some other way that's going to lower the quality of treatment? ... Will he do what I want but resent it and not be quite as good ... or in some way ... [be] detrimental to my quality of care?"

Many reported feeling cast "in the role of supplicants, operating under the premise that 'doctor knows best.'" Avoiding asking too many questions or disagreeing with advice, they admitted adopting deferential attitudes "to avoid 'displeasing' or 'disappointing' their physician." Patients may instinctively worry whether doctors, just like others whose prestige begets power, are liable to revert to dominant behaviors if they perceive their status is being challenged.

Fear of physician retribution may—at least occasionally—be well-founded. In 2012, a report conducted by the Patients Association in the UK—an independent charity—concluded, "One of the great unspoken scandals of the NHS [National Health Service] is the risk of being removed from your GP's list because you make a complaint against them."[70] Almost 1 in every 20 individuals who submitted a formal complaint about their family doctor ended up being deregistered, and more than a fifth of all complaints about GPs to the UK's Parliamentary and Health Service Ombudsman related to "inappropriate deregistration."

Responding, and speaking on behalf of the UK's 50,000 family physicians, Dr. Clare Gerada, former Chairperson of the Royal College of GPs, stated, "We are really disappointed."[70] Regrettably, this was not a prelude to a public apology; Gerada elaborated: "We would have preferred the Patients Association to work with us to find ways of supporting GPs ... rather than unfairly criticizing GPs as this report does," adding, "GPs actively encourage patient feedback and take concerns very seriously."

Although we must sympathize with the medical profession as an overstretched and burned-out workforce, medicine must be open to the possibility of failures in patient care. Low numbers of complaints might not tell the whole story. A systematic review of 68 surveys reveals that the biggest barrier to patients speaking up about medical errors is feeling inferior to doctors.[71] Researchers concluded: "The hierarchical, elitist and paternalistic culture of the medical profession was often a barrier to patients' willingness to engage with their safety. ... It was seen as inappropriate to challenge clinicians ... some patients were fearful of questioning medical authority ... while others were afraid of being rude to or offending the doctor." Because of "concerns about being labelled 'difficult' and clinicians responding negatively or defensively to being questioned" patients believed that they were "*actively* protecting their personal safety by assuming a relatively *passive* role."

Taking a surgical retractor to patient feedback, in 2019, UK research from the Care Quality Commission found that almost 7 million people in England who accessed health or social care services in the previous five years had concerns about their care but failed to complain; over a third felt that nothing would change as a result.[72] In the UK in 2020, research conducted by the Parliamentary and Health Service Ombudsman found that one in five people did not feel safe in the mental health service that treated them; however, nearly half said they were unlikely to complain, and one in three doubted their complaint would be taken seriously. Explaining their reasons, 25 percent were "concerned it would affect how I was treated."[73]

Having reservations about raising a hullabaloo is not just a character trait of the famously buttoned-up Brits. Many Americans are cautious of clinicians too. In 2019, in a survey of more than 10,000 patients across eight hospitals, nearly half experienced a problem during hospitalization but around a third were uncomfortable about speaking up.[74] Patients who were already "low status"—the elderly, people with worse overall health including mental health, and those who didn't speak English at home—were more reticent to vocalize complaints.

Tom Lawson is a highly accomplished academic. He is also a person with street smarts. Dr. Parker was respectful, caring, and courteous.

However, even in the best of circumstances, Tom succumbed to a version of "doctor deference." He admitted: "It completely escaped me that I was being diffident around my doctor." As if confirming the problem he added: "You know, being more assertive might have helped. But only if the doctor wasn't threatened or annoyed by it."

Although social scientists have long lamented doctor-centered power dynamics, it is important to add that doctors' comportment is not premeditated or deliberate. What we've explored in this chapter are human psychological tendencies. Nor do insights into the origins of status offer an excuse for medical (mis)conduct. Human behavior is not rigidly fixed or immutable. On the other hand, to change the world, we need to understand it and our proclivities within it.

Metaphorically speaking, doctors need to come down from their pedestals, and patients need to get off their knees. However, the sheer pressures and stress in our time-limited clinic visits make it harder for both parties to take a step back, and deliberately steer and engineer their reactions. This makes it difficult—in *Dr. Phil* parlance—for doctors and patients to become the "best versions of themselves."

Could technology have strengthened Tom Lawson's voice, helping him to avoid delays, embarrassment, and fear of doctor-bothering? In the next chapter we'll explore whether digital tools can curb cap-doffing in the clinic.

5

Pouring Our Hearts Out to Machines

"You know doctor, I really like this computer better than the physicians upstairs."

So proclaimed the very first person to "talk" to a computer about their health.[1] The year is 1966, the location is a hospital at the University of Wisconsin,[2] and the physician is Warner Slack (1933–2018). On hearing the patient's candid admission, Dr. Slack is not insulted; he is relieved. Prior to launching the study, Slack's colleagues were deeply suspicious about the idea of computers as medical interviewers.

"A number of colleagues expressed concern over our idea," he later recalled.[1] "Some people were saying it could not be done. Some felt it *could* be done but *shouldn't* be done, and others said patients would not permit themselves to be interviewed by a machine."

"To be perfectly frank," Slack confessed, "it took quite a while to get up the courage to try the first clinical interview."

Putting time-honored patient medical-history-taking to the test, Slack recruited 50 older patients. He requested they give detailed interviews about their allergies to an early "LINC" or "Laboratory INstrument Computer". Sitting at the machine—monster-sized by today's standards—patients followed a series of pre-programmed questions. Depending on how patients answered, the questionnaire branched out to pose new queries. Although the computer program was limited, and responses were restricted to "yes," "no," "maybe," or "don't understand," the set-up was straightforward.

As the first interview—between an older gentleman and the computer—got underway, Slack recalled something curious: "[I]t became clear there was a rapport between him and the machine. He was laughing out loud at some of the things from the computer. Some I had intended to be funny; some I hadn't. He was saying things like, *'That was a dumb question—you already asked me that!'* "[1]

Rather than lamenting the shortcomings of the program, Dr. Warner Slack recognized something unique was unfolding: "He would never have said that to me as a doctor with white coat and Bakelite nametag."

Slack later reflected in an academic article: "The physician presents an authoritarian figure," and yet "the patient was very comfortable with the machine and criticized it freely."[1]

Proving great minds think alike, the very same year (1966) Joseph Weizenbaum, a German American computer scientist working at Massachusetts Institute of Technology, Boston, created the world's first chatbot—a computer program aimed at simulating human dialogue. Weizenbaum named the program after Eliza Doolittle, the working-class heroine in *Pygmalion* who learned to pass herself off as upper class—though in this case, Eliza's job was passing "herself" off as a person. Here is a sample conversation:[3]

Human: Men are all alike.
ELIZA: IN WHAT WAY?
Human: They're always bugging us about something or other.
ELIZA: CAN YOU THINK OF A SPECIFIC EXAMPLE?
Human: Well, my boyfriend made me come here.
ELIZA: YOUR BOYFRIEND MADE YOU COME HERE
Human: He says I'm depressed much of the time.
ELIZA: I AM SORRY TO HEAR YOU ARE DEPRESSED

The program served as a live demonstration of the renowned Turing Test—devised by mathematician and World War II Enigma codebreaker Alan Turing—which challenges people to determine whether they are talking to a human or a machine. Inviting students and colleagues to test the program, Weizenbaum was astounded by the intimacy forged between humans and the machine, culminating in a request by his secretary that he leave the room so she might talk with the computer in private.

Where Joseph Weizenbaum was disturbed by the possibility of deception and the displacement of human intimacy, Warner Slack saw medical potential. Like a Martin Luther of machines in medicine, in 1966 he nailed his theses to the cathedral door of clinical practice—in the top journal, *The New England Journal of Medicine*. In a landmark publication entitled "A computer-based medical-history system," he dared to slaughter a sacred clinical cow—to question whether patients might open up more with computers than with clinicians.[2]

Computers versus clinician interviewers

"The one distinction I have in the world—hands down—is I've put more people in front of a computer before they were seen by a doctor than anyone else."

These are the words of Dr. Allen Wenner, a retired primary care physician from Lexington, South Carolina. Like his friend and mentor Dr. Warner Slack, Wenner saw a role for computers as information gatherers and wanted to see them in action.

Before they set foot in his office Dr. Wenner usually knew why his patients wanted to see him. Wenner wasn't some sort of clinician-clairvoyant. Instead, his patients used an online tool to disclose their problems prior to their visit. Using a secure website that he devised called Instant Medical History, linked to his health center, patients answered a series of questions explaining why they wanted medical attention.

Dr. Wenner spoke to me from his bookcase-lined home office and expressed a strong conviction about the value of the product. He told me how it works: "The website opens with a free-text entry. So, if you put in sore throat, it'll ask you a set of questions with closed-ended [e.g., "yes"/ "no"] responses. Depending on the problem, the tool pulls up a specific clinical questionnaire allowing patients to enter in as much information as they want."

"By the way," Dr. Wenner adds, "if you've five different problems, you can add them in. Patients could choose to spend a couple of minutes, or even half an hour, answering questions online. When they've had enough, they can stop."

"A patient might have a cough," he explains. "You can go onto Instant Medical History and type in 'cough.' And it'll say something like, 'Is it worse at nighttime when you're lying down, or is it worse during the day?' And you might think, 'Oh yeah, actually it's worse at night.' 'Is it worse going upstairs?' It leads you through questions, so that you've thought about what kind of cough you've got."

"It's not like going into the doctor and saying, 'I've got a cough', and something important could be missed."

This concern is not a small one. Take the UK, where signage across many of the UK's National Health Service (NHS) primary care waiting rooms encourages patients to be specific and efficient in their visits: "1 Appointment, 1 Problem, 1 Patient = 10 Minutes." The problem is that patients may not know the priority issue or may feel unsure about how much detail to share.

After logging their problems in Instant Medical History, patients have a chance to first read through their own report and then download it for their own records. With a warm southern drawl, Dr. Wenner took me back to the beginning: "I knew about Warner Slack's work. And the eureka moment came when I realized computer interviews could improve the quality of care. Twenty years ago, I did something about it."

"I did a study of discharge diagnoses of people leaving my office. I looked at every patient with hiatal hernia [a condition where the stomach bulges into the diaphragm]. It's a condition that causes gastroesophageal reflux—or heartburn."

"If you go to the medical texts, there are around 11 questions you can ask related to this condition. Well, I found that on Mondays I asked seven questions, compared with Friday afternoons when the average was 4.5."

"That isn't really a study you want to publish on yourself: *'I'm only half as good on Friday afternoon than I am on Monday.'* But the numbers showed this was a quality issue."

During appointments, data entry compounds the pressure on the visit. "Death by 1,000 clicks" is a common headline in medical blogs, and time-and-motion studies show the extent of the problem. In 2020, an analysis of log entries from 100 million patient encounters by more than 150,000 doctors working in outpatient care revealed they spent an average of 16 minutes and 14 seconds using the electronic health record for every person.[4] Given that there are fewer doctors today and the average appointment slot in the US is around 15 minutes, doctors are clearly working on negative time.

"Mostly, updating patients' electronic medical records is done after the fact rather than in real time in the office," he told me. "And often they're not done immediately after the visit but in the evening or on weekends. Doctors call it 'date night with Epic' [Epic is the electronic medical records system used by around half of all American physicians]. I hate to tell you," Dr. Wenner explains, "if you're seeing someone on Monday and updating their record on Saturday, the history could be so inaccurate as to be worthless."

Warner Slack once opined that patients are the "largest and least used resource in healthcare."[5] Nobody is more invested in the accuracy of a health record than the patient for whom it is written. Errors in records, says Dr. Wenner, are still taboo among clinicians: "When patients enter their own health information prior to the visit, this saves doctors an average of 57 clicks per visit. *That's huge."*

Dr. Wenner tells me: "I actively wanted my patients to contribute to their records so that I got a true representation in their words. Even if the documentation wasn't completed, it was started." Warming to the theme, he becomes blunter: "Look, some complaints doctors don't want to deal with—headaches, weakness, patients who are tired or dizzy, patients with back pain. ... Doctors won't admit it, but nobody wants to deal with those kinds of ailments. With this tool, patients can invest great detail in describing exactly these problems."

Computers are incapable of being bored by patients' symptoms and with Allen Wenner's pre-visit tool, medical information-gathering was becoming automated.

Computers as conversationalists

More than likely, you have not used software such as Instant Medical History, and Dr. Wenner tells me with a trace of weariness: "I will say that doing this for 30 years has not been the most profitable thing I've ever done."

Computer-mediated interviews never made the healthcare hit parade. There are several reasons for this. One major problem is doctors never saw the use for it—of which more, later. Another is that, until recently, most computer "interviewers" on the market made for rather rigid "conversationalists." This is because they have primarily relied on what computer scientists call "expert systems." Let's take a closer look at the anatomy of this artificial intelligence.

In the history of AI, expert systems emerged in the 1970s and 1980s, and these early computer programs were designed to emulate the decision-making of human experts in specific domains. In medicine, a bunch of doctors might gather in a room to figure out what kind of knowledge needed to be programmed to ask patients about, say, whooping cough. This information would then be structured into decision trees within computer systems, aiming to capture the essential data points associated with the illness.

Because they principally rely on numerous pre-programmed questionnaires, however, these systems are laborious for programmers to maintain. They are also dependent on highly specific input—words and phrases entered by the patient—and this means they make for inflexible and infuriating interfaces. At the time of writing, most healthcare chatbots still primarily rely on expert systems.[6] These are the same systems that underpin many customer service phone lines. As any consumer who has "conversed" with such services knows very well, automated phone lines are highly irritating. In a Dantesque descent into hell, they can lack nuance, are repetitive, and very obviously fail to "understand" what we're telling them.

This is partly why devices like Instant Medical History never made it big. Too many closed-ended questions to wade through, repeated requests for information, and the humdrum nature of the formats make them conversationally clunky.

Still, in many ways, Dr. Allen Wenner's story is a lesser-known success. Implementing AI is not a matter of sprinkling digital stardust on doctors: without a clear rationale for *why*, *when*, and *how* to use computers in communication, these tools can exacerbate doctors' workflow problems.[7] Dr. Wenner envisioned computer interviews merely as a pre-appointment tool for gathering thorough, candid patient histories—not for triage, diagnosis, or outright replacing in-person visits. He did not simply bolt these systems onto care and hope for the best. He gave his patients and his colleagues proper guidance to manage their expectations, explaining what the systems are good for and how they work. This contrasts with the rushed roll-out of digital tools like NHS England's "eConsult," which, particularly during the pandemic, left patients and doctors frustrated. Inappropriate implementation of these tools created unrealistic expectations, undermining their usefulness.

Unlike older "expert" AI, newer chatbots, like ChatGPT, Claude, Gemini, and DeepSeek, have revolutionized AI by being much more "conversational" and engaging. These chatbots largely use "connectionist" AI, which attempts to mimic how the human brain learns and makes decisions by adjusting millions—sometimes even billions—of internal settings (or "weights") as they process data. Powered by advanced neural networks trained on vast amounts of text data, these chatbots can identify patterns, make predictions, and continually improve. Known as "large language models" (LLMs), they are built on what is called a transformer architecture, which allows them to analyze text and word relationships with exceptional precision to produce humanlike conversations. Demonstrating their sophistication, ChatGPT made history in June 2024 as the first AI reported to pass the Turing Test, consistently convincing people they were communicating with a human rather than a machine.

It is still early days, but emerging studies suggest that these newer AI tools might function as exceptional clinical interviewers. In 2024, a team at Google unveiled a new LLM chatbot called "AMIE"—short for Articulate Medical Intelligence Explorer.[8] In a trial comparing the bot to 20 board-certified primary care doctors using 149 clinical case studies, the team employed patient actors, and simulated medical conversations for a variety of different complaints. "Patient" confederates engaged in typed dialogue with either actual primary care physicians or AMIE.

In blinded assessments, panels of patients and specialists judged AMIE to be significantly better than doctors on multiple measures of history-taking—from politeness to trustworthiness, in terms of listening and explaining the condition, and from being honest to involving patients in treatment decisions. In total, patients rated the bot significantly superior to doctors on 24 out of 26 measures, and equivalent for the remaining two. Meanwhile, specialist physicians rated it superior on 28 out of 32 evaluation measures.

This is only one study, though we can expect many more. But it signals, as Drs. Warner Slack and Allen Wenner long recognized, that, compared with human clinicians, computers have extraordinary potential to put patients at ease.

Computers as clinical confession boxes

Recall that 93-year-old Tom Lawson delayed discussing his peeing problem with his doctor out of embarrassment. A substantial yet frequently overlooked body of research shows patients are more likely to share their clinical and social problems with computers compared with human clinicians. This suggests that, when seamlessly integrated into the patient's journey, technology might well have turned Tom from a cap-doffing supplicant into a more candid confessor of his symptoms.

Take a review published in 2003 entitled "The patient–computer interview: a neglected tool that can aid the clinician," which concluded that computer-based interviews generated between 35 to 56 percent more detail than face-to-face visits, with most of the information gathered identified as new and clinically important.[9] For example, a study in Germany found computer-histories generated an average of 3.5 new problems per patient—information that had not been elicited from clinic visits.[10] Researchers concluded that computer interviews were "a far superior method for collecting historical data than the physician interview alone."

Dr. Allen Wenner described his first-hand experiences with sensitive patient testimonies: "A provider will ask, 'How much do you drink?' The patient might say, 'Five a day.' The doctor scrunches his face. The patient says to themselves, 'I better not go there.'" He adds, "When you take away that body language and put patients in front of a machine it's easier."

Patients are indeed less likely to lie on online questionnaires than with doctors. Studies show they are more candid about booze consumption,[11] tobacco use,[12] illicit drug habits,[12] HIV risks,[13] suicide attempts,[14] and even problems at work.[15] Adults and adolescents disclose more about their sexual behavior, impotency, or other sensitive gynecological worries.[13,16]

"Another one is domestic violence," Dr. Wenner emphasizes. In the United States, one in four women and one in nine men experience severe intimate partner violence.[17] "We had seen only two cases of domestic violence the year before John Bachman [Professor of Medicine at the Mayo Clinic, Rochester] and I made up a questionnaire scale. This was in the 1990s. We constructed four questions and put them into the online tool."

After implementing the tool, the volume of personal disclosures was overwhelming: "We had so many cases of domestic violence in my practice in South Carolina that we had to find another clinician to help deal with the problem. We didn't know who to refer all these cases to ... the numbers showed what we were really missing."

In 2015, a review of randomized controlled trials found women were 37 percent more likely to disclose intimate partner violence to a computer compared to a health professional during a face-to-face visit.[18] In 2017, a literature review of intimate partner violence reported that computer-based screening methods allowed the victim to "answer various questions without being interrupted or without the feeling of being judged and embarrassed."[19]

Even the most rudimentary of electronic communication—email—can help patients say more. Health communications expert Dr. Debra Roter of Johns Hopkins University conducted one of the first studies exploring the content of emails sent between patients and doctors.[20] While physicians dominated the volume of dialogue during office visits, Roter and her team found patients expressed twice the number of ideas online, using double the word count. Similarly, in Denmark, by 2019, more than one in five primary care consultations were conducted via online messaging, with research concluding that "email consultations make it easier for patients to write than talk about sensitive or emotional topics, suggest their own hypotheses for discussion and ask questions."[21] In the US another study found that patients offered secure email with their doctors were 32 percent more likely to get a mammogram, 11 percent more likely to undergo a Pap test, and were 55 percent more likely to receive colon cancer screening.[22]

Why patients are mouthier with machines

To revolutionize healthcare, tech innovators must first understand the specific pain points and challenges embedded in traditional systems. This book argues that creating effective digital tools requires more than innovation—among other things, it also demands a deep awareness of when and why patients might engage more openly with machines. By addressing these nuances, we can improve the design of digital tools and ensure their better integration into real-world healthcare, ultimately enhancing patient outcomes.

Let's review what we know. Warner Slack witnessed patients openly disclose their symptoms to his LINC machine while occasionally ridiculing the computer as a lousy conversationalist ("told you that already!"). While people appear to pour their hearts out to machines, they may also get fed up with too many repetitive or clunky questions that "stubbornly" fail to "grasp" what they are saying. Indeed, people mouth off at technology. Survey research reveals that one in three of us have "abused" our computers—whether by cursing, yelling, or straight-up striking the poor machine in frustration.[23] Even more fascinating, the more frequently it malfunctions the more we're inclined to say our computer behaves like it "has a mind of its own" with "its own beliefs and desires."[24]

To understand why we can be more expressive (and scathing) in front of technology compared with doctors, and how we can harness these human responses for optimal medical ends, it is valuable to explore when and why we treat objects, even momentarily, like real people.

Personifying objects is as old as human history. In the sixth century BC, Xenophanes coined the term "anthropomorphism" after observing humanity's uncanny tendency to depict gods in their own image. Scottish Enlightenment philosopher David Hume also wrote about the quirk: "There is a universal tendency among mankind to conceive all beings like themselves, and to transfer to every object, those qualities, with which they are intimately conscious."[25]

In the modern era, one example of anthropomorphism attracted the amusement of the world's media. In October 2022, amid newly appointed Liz Truss's turbulent premiership, the shortest of any UK prime minister in history, British tabloid newspaper *The Daily Star* streamed a "Live Lettuce Cam" on YouTube, playfully asking: can Liz Truss outlast a 60p supermarket lettuce?[26] Pitting the PM against "Liz Lettuce" proved a viral hit but viewer interest only surged when the vegetable was gradually humanized—first adorned with a pair of googly eyes, soon after with a mouth and wig, and occasionally snacking on English pub favorite pork scratchings. Marking the vegetable's eventual triumph over Truss, an image of the lettuce was beamed across the Houses of Parliament. "Lizzy Lettuce" was a stroke of tabloid genius because it engaged readers' and the public's attention.

We might wonder if we simply perceive visages in objects because they're *similar* to faces, but this begs the question about *why* we detect similarity in the first place. Taking a longer view of human nature, perceiving and recognizing human faces was adaptive.[27] Better to see a face—especially a predator's—than not. Promiscuity in how our brains process stimuli associated with faces means we sometimes see them where they don't exist—in an iceberg lettuce, on toast, in tea leaves, or floating in the clouds.[28,29] From surprised-looking plug sockets to big-nosed doorknobs, we are impressively attuned to detect faces in inanimate objects.

Just like *The Daily Star*'s editor, savvy designers and astute advertisers exploit this anthropomorphic tendency in product development and marketing. When it comes to brand icons, the more humanized the product symbol, the more likely it is to be adopted by advertisers and command lucrative sales.[30] Consumers perceive upturned grilles on cars as friendly, and slanted headlights as aggressive, with vehicles exhibiting both expressions (displaying a confident car persona?) most preferred by petrolheads.[31] And those drivers who perceive their car as having a "warm" personality are less likely to replace the vehicle or send it to the scrap heap.[32]

Aside from face detection, humans intuitively "read" behavior in terms of beliefs, desires, and intentions: often referred to as "theory of mind." This evolved capacity renders behavior understandable and predictable. Our natural tendency to "see" agency in abstract objects is exquisitely demonstrated in a 1944 study entitled "An experimental study of apparent behavior."[33] The experiment encompasses footage, devised by psychologists Fritz Heider and Marianne Simmel, of two-dimensional shapes—a small triangle, a larger triangle, and a circle—moving around the screen. Check it out on YouTube.[34] Watching the patterns and timing of movement, viewers can't help but see the objects as displaying social behavior, complete with emotions and a sense of purpose: the small triangle and circle appear involved in a romantic tryst, with the aggressive large triangle jealously hell-bent on splitting them up.

We blame the stray object that seems to trip us up, swear at smoke alarms, or wallop the photocopier that lets us down. While there are individual differences in the extent to which individuals anthropomorphize, people tend to prefer robots that express themselves in a more humanlike manner.[35] Anthropomorphized robots are perceived as more intelligent, more attractive, and more credible than non-anthropomorphized ones.[36,37] Neuroimaging shows that our brains process human actions and those of anthropomorphized robots using the same neural circuits.[38]

These examples offer hints at how we might use digital interfaces to avoid some interfering human traits—such as cues of intimidating prestige or dominance—while ramping up others, for example, warmth and friendliness. In short, design choices in technology can trigger anthropomorphic responses, shaping what we disclose about our health, with significant practical and ethical implications.

Research shows that the more a computer interface stimulates facial recognition or theory of mind, the more likely we are to interact as if in the company of a person.[39,40] In a study led by Lee Sproull at Boston University, participants were more likely to present themselves in a positive light when responding to a human face displayed on a screen, compared with a text-display questionnaire.[40]

Some research suggests that embodied agents such as avatars—cartoonlike images on the screen—might do a better job of helping patients to disclose sensitive information.[41] Gale Lucas and her colleagues at the University of Southern California found people were looser-lipped and visibly more expressive when talking with avatars they believed were operated by computers rather than by people.[42] Exiting the interview, researchers were struck by the unsolicited feedback voiced by the study participants; for example: "This is way better than talking to a person, I don't really feel comfortable talking about personal stuff to other people," said one interviewee. "A human being would be judgmental," opined another, "I shared a lot of personal things, and it was because of that."[42]

The features of the avatar—its clothing, appearance, speaking voice and fluency, and the quality of the animation—likely influence our responses. Perhaps avatars expressing warm, friendly, nonthreatening faces extract more candid, less inhibited clinical confessions.[43]

Dr. Ryan Schuetzler is an Associate Professor at Brigham Young University's Marriott School of Business, where he specializes in chatbots and human behavior. In a series of experiments, he also found that people disclosed more about alcohol intake and health behavior when interviewed by conversational agents than with face-to-face human interviewers.[44] However, conversational agents that were "better"—that is, judged as more humanlike in their communication—

elicited more socially desirable and less truthful responses. The more the conversational agent seemed to understand, the more people actively tried to deceive it.[45]

This research confirms that replicating our regular interactions with humans can sometimes be a mistake. Again, how we design these tools is likely to influence what we say. Perhaps if the interface is perceived as exceptionally approachable, attentive, friendly, and safe, this could supercharge trustful relationships in ways that lead us to open up. Given the right cues, people might perceive these tools as trustworthy clinical confidants. Hinting at this, in their study "When the bot walks the talk," researchers at the University of Basel, Switzerland found that user trust of AI hinged on key factors including the chatbot's perceived competence, integrity, and human-like qualities.[46]

Other studies suggest there could indeed be some supercharged AI exceptionalism going on, with growing evidence that many people are beginning to rely on LLM chatbots for their health needs. Take the 2024 UserTesting survey we mentioned in the last chapter, where more than half of respondents said they had used chatbots such as ChatGPT to investigate their medical symptoms.[47] Asked why, 51 percent said they did so because they were embarrassed by what they were experiencing.

It could also be that the more we use them, the more comfortable and familiar these tools become and the more we depend on them. In psychology the "mere-exposure effect" describes the scenario where people develop a preference for things merely because they are familiar with them.[48,49] The more we are exposed to something, and the experience is positive, the more likely we are to trust and like it. Several theories have been proposed to explain the evolutionary origins of this phenomenon. For example, people might prefer familiar stimuli because they're indicative of a safer, more predictable environment. Positive feelings associated with familiarity may have been adaptive in social bonding: being exposed to the same individuals or groups over time could foster trust and cooperation, essential for survival in communal living situations.

So, how can we leverage AI to encourage patients to be open and honest when it matters most? Health tech innovators might strive for a kind of Goldilocks of anthropomorphism: interfaces that bypass the psychology of status, and avoid the social anxieties associated with human judgment, and yet feel exceptionally personable, friendly, trustworthy, and natural enough to talk to. Yet in the wrong hands, these tools could exploit us, something we'll explore later in the book. Navigating and balancing these concerns in healthcare demands both ingenuity and a deep sensitivity to the vulnerabilities of our Stone Age psychology.

Unmasking our medical secrets

"My body was nervous," says Tom Lawson. "I didn't think I was nervous. But they would do my blood pressure reading in the clinic and say, 'Oh, it's 160/90.' Then they'd do it again and it would be a little better, but still too high."

If traditional visits can distort how we present our symptoms and signs, beyond chatbots, a range of surreptitious technological tools can do a better job of unmasking our hidden health secrets. Take Tom's blood pressure reading. Stress in medical encounters can elevate patient measurements, causing what's known as "white-coat hypertension." This phenomenon can be a serious issue: ascribing high blood pressure to clinic anxiety could camouflage real health complaints.[50] "When I do it at home," says Tom, "with a special cuff, the reading is usually around 110/60."

Latest data from the Organisation for Economic Co-operation and Development show that people in the US and the UK see their doctor fewer than three times per year for an average of only 30 minutes in total.[51] It is little wonder that doctors operating in traditional visits lack detailed pictures of our signs and symptoms. Remote monitoring joins the dots. Via electronic blood pressure cuffs, heart rate monitors, and a host of other gadgets this approach helps patients and doctors obtain a more reliable representation of health status, cutting out the need for intermittent clinic confessionals. Our health data can be relayed instantly, both to the patient and their clinical team.

Dr. Sara Riggare, a digital health researcher and a colleague of mine at Uppsala University in Sweden, uses her experience of living with Parkinson's disease to highlight healthcare's time imbalance. She emphasizes that each year she spends just one hour with her neurologist, dedicating the remaining 8,765 hours to self-care. Sara showcases the importance of patient self-empowerment through self-management and health-tracking.

Research supports Sara's insights. For example, combining the continuous data-gathering of the sort adopted by Tom with the video visits used by his daughter Jen provides a powerful method of managing chronic illnesses such as type 2 diabetes. In 2015 a Cochrane review of 93 trials concluded there were numerous advantages of this approach over face-to-face appointments.[52] In 2021, a systematic review of 91 studies concluded that remote monitoring reduced acute hospital readmissions for patients with cardiovascular disease and the range of chronic lung conditions known as "COPD" or chronic obstructive pulmonary disease.[53]

The small rectangular artifact we juggle between our hands and back pocket is also passively privy to salient health information. From how much time we spend in the pub to what we spend on healthy food, from how often we talk to friends and how isolated we may be to how frequently we visit green spaces—our smartphones are privy to a lot. Beyond what we choose to disclose—either to doctors or via computer-mediated questionnaires—a near-constant stream of digital activity captures a wealth of clinically salient information.

Come rain, shine, or the most challenging bowel movement—most of us even admit using it on the throne—there will be social-media-liking, news-swiping, and fastidious message-checking. In Western Europe, people spend about 2.4 hours per day on their cell phones.[54] In the US, the figure is more than double. In 2024, the average American spent 5 hours and 1 minute daily on their phone—adding up to 2.5 months (or 76.3 days) per year of nonstop scrolling and thumb action.[55] A Pew Research Center survey reported that nearly half of teens in the US used the internet "almost constantly."[56]

Mainstream medicine doesn't currently tap this digital stream for the good of its patients. However, integrating our personal information derived from our devices with a richer understanding of each patient's daily life could better inform our understanding about illness. Dr. John Torous' clinic is a rare exception. John, a psychiatrist and colleague, serves as the Director of Digital Psychiatry at Beth Israel Deaconess Medical Center at Harvard Medical School, where I'm also an affiliated researcher. He sat down with me to discuss what smartphones can do that doctors can't.

"The fact is," he says with singular humility, "99.9 percent of the time doctors do not understand what's going on in each patient's life."

"Illnesses are complex and multi-factored. There's a genetic component to illness but there's also a huge environmental component. It's been very hard, classically, across all of healthcare, to understand the environmental component of illness: What happens when you're outside the doctor's office? What really happens with patients' sleep or how much they are exercising?"

As we saw in the last chapter, patients often finesse facts or misremember behavior between visits, but with smartphones or smartwatches as constant companions, such deceptions—intentional or not—become more difficult. "Digital phenotyping" is the term for information derived from consumer electronic devices to yield information about health. Broadly, the idea is that such data can, if carefully employed, offer insights into the user's health and wellbeing. For example, tweets in the early hours can signal insomnia; in contrast, regular and predictable patterns of phone activity can be a proxy for a strong daily routine. Not just what we say, but how we say it, might offer clues about our health.

John offers an example: "We've always known it's healthy to be outdoors. So, in a study we published, we explored this using phone data. With full permission, we received a ping from patients' phone GPS [Global Positioning System]. We didn't look up exactly where they were. Instead, we converted the location to an index of green space, based on state maps. We could then gauge, over a month, how much green space exposure they had, versus how much urban, concrete environment."

The digital psychiatry team sent patients very brief online questionnaires, again via their phones, to record, in real time and in their own surroundings, how anxious or depressed they were feeling. The team tracked and compared each patient's environment with their symptom reports.

"What we found was an association; we can't demonstrate a causal relationship here—as the saying goes, 'correlation isn't causation.' But we showed there *is* an association with patients' mental health." Green space was associated with better symptom scores for depression, anxiety, psychosis, sleep, and sociability.[57]

Dr. Torous harnesses this approach to improve the quality of his care. His team invites patients to download mindLAMP, a free, open-source digital app devised by John and his colleagues, onto their phones—the same platform he used to explore exposure to green space. With each patient's permission, mindLAMP encrypts and sends log-data from the user's phone to a secure healthcare server for analysis. The app declares up front what it can collect, and users must consent to each sensor which can potentially gather different kinds of information from their phone: "This data is treated with utmost security and the app is not commercial," he keenly underscores. "It's not sending your data to Google Analytics. It's not partnering with anyone other than their secure health server and the medical team."

"We bring this data to the table during visits and look at the graphs together. By quantifying the relationship between their activities and their wellbeing, we can put it to patients that *this* could be making a real difference in their life." He adds: "It might then be a case of saying, 'Hey, we seem to see improvement, it's subtle, let's keep going and not switch to a medication yet.' At the very least, the data will start a conversation."

Some patients are simply not interested in using smartphones in the clinic. A separate worry is creating phone-induced digital dependencies. This, in turn, might encourage people to stray onto social media which could potentially lead to the kinds of unhealthy social comparisons that chip away at wellbeing.[58,59] Still, many patients are enthusiastic and this is reflected in John's growing waiting list.

Reliable data and its responsible use are also crucial for accurate health insights. At the very least, digital phenotyping depends on our phones being switched on, juiced up, and with us; as John says, "Low power mode on phones is kryptonite for data collection." Phones are also far from perfect when it comes to accurately tracking our behavior. Something as simple as sharing a phone with a family member or leaving it at home can easily distort the data, highlighting the importance of context in interpreting health insights.

Despite their fallibility, our phones and internet-enabled devices can collect and fuel significant information on us. For example—and again, with the full consent of participants—studies in Denmark investigated distinctive voice features of patients with bipolar disorder. Over an extensive period of more than two and a half years, a Danish team analyzed participants' phone calls to reveal distinct vocal markers associated with depression, and the extreme highs of mania.[60] They found that the number of text messages and duration of phone calls among patients with bipolar disorder significantly differed from those of healthy participants, with an increase in phone use prior to manic episodes.[61]

Covert clues embedded in our online dispatches on Reddit, Facebook, or X/Twitter can also reveal more than we might intend. Psychologists have long recognized that there are linguistic markers associated with our moods. People who are sad or depressed tend to pepper conversations with more first-person pronouns ("I", "me") and fewer first-person plural pronouns ("we," "our"), and use more negative emotion words, adopting fewer positive words.[62]

Using content from Facebook posts, a team led by Johannes Eichstaedt, Director of the Computational Psychology and Well-Being Lab at Stanford University, analyzed securely anonymized data from 1,200 participants who agreed to provide access to both their Facebook status updates and electronic medical records.[63] They found that linguistic markers could predict depression with "significant" accuracy, up to three months before doctors provided a diagnosis. More recently, research shows that AI models like ChatGPT can detect early dementia by analyzing subtle speech patterns—hinting at a future where AI aids early diagnosis with remarkable accuracy.[64]

Other studies have excavated the medical secrets embedded in our electronic health records. Many of these archival digs unearth highly stigmatized health and social problems.

For example, using historical data from medical charts combined with machine learning models, a team from Harvard Medical School predicted cases of domestic violence with high precision.[65] The model picked up on deeply buried but distinctive patterns of signs and symptoms. Moreover, machines were much more perspicacious than doctors, making reliable risk assessments, on average, 10 to 30 months in advance of human experts.

Suicide is the twelfth leading cause of death in the United States, and the third leading cause of mortality among individuals aged 15 to 24.[66] Approximately 50 percent of adults who die by suicide visit a healthcare professional in the four weeks before their death.[67] Yet doctors' predictions of which patients are most at risk are barely better than random guessing.[68]

DISCLOSING OUR SYMPTOMS

Mining the medical secrets buried in our charts, AI could help prevent self-harm or even suicide.[69] Researchers used a prediction model trained on the records of more than 3.7 million patients, across five health centers, to forecast future risk of suicidal behavior.[70] The model accurately predicted nearly 40 percent of suicide attempts on average 2.1 years before they occurred. No single predictor was identified—instead, a powerful gestalt of factors amplified the signal.

Clearly, how health organizations, tech companies, and other institutions collect and share our information raises important privacy concerns. Already you may feel uneasy about medical charts being mined for health risk scores. Reflecting on the potential to identify suicidality, one of the study investigators Dr. Ben Reis sees clear benefits: "We envision a system that could tell the doctor, 'of all your patients, these three fall into a high-risk category. Take a few extra minutes to speak with them.'"[71] AI could potentially facilitate deeper patient–doctor revelations and connections.

Computers could have loosened Tom Lawson's tongue, freeing him from diffidence and embarrassment in front of his doctors. In a wide variety of ways, technology has the potential not only to automate medical interviews but also to innovate—radically changing how we gather clinically relevant patient information. With echoes of Weizenbaum's worries, however, patients are also in danger of oversharing: cold hard machines serve as cunning collectors of our medical secrets.

PART III
DIAGNOSIS AND TREATMENT

6

Crafting Clinicians

In her heartfelt 2019 Oscar acceptance speech, the globally famous and Grammy Award-winning singer Lady Gaga dispatched rousing advice to the aspirant stars of tomorrow. "If you're at home and you're sitting on your couch," she counseled, "and you are watching this right now, all I have to say is that this is hard work. I've worked hard for a long time and it's not about winning . . . it's about not giving up."

The singer's advice echoes basketball legend and billionaire Michael Jordan's top tips for achievement: "If you do the work, you get rewarded" and "The more you sweat in practice the less you bleed in battle." From Apple's Tim Cook to former Starbucks chief Howard Schultz, business magnates counsel that early starts and late nights—a commitment to punching in the hours—is the secret sauce of success.

Medicine also celebrates its "all-stars." Gongs are annually awarded to the world's finest physicians at in-house gala events. Nominees for the "Top Doctor Awards" are picked from among top-flight physicians in the US and Europe. America's "Above and Beyond Award" is presented to the pediatrician whose "dedication, talents and skills have improved the lives of countless childhood patients."

Medical decoration ceremonies might strike us as somewhat decadent affairs ("This award is dedicated to all my patients who didn't make it"). However, the idea of a "good doctor" or "top specialist" is familiar. As patients, we want to be treated by "the best," an aspiration that hints not every physician is top rung—after all, 50 percent of doctors graduated in the bottom half of their class.

Imagine, however, striking the medical jackpot with your very own Lady Gaga or Michael Jordan of medicine—let's call her Dr. Amazeballs. Diligent, dedicated, and devoted to her patients, she spends her spare time reading up on the latest medical research. Amazeballs MD is unfailingly empathetic but also cool-headed. She seems like a dream come true and, unfortunately, she is just that—a mere physician fantasy figure.

Medicine is tougher and messier than most fields. It is also resistant to perfection—no matter how much effort is put in. Clinical practice is drowning in "noise"[1]—from the time of day a judgment is made, to how professionals are trained, and the idiosyncratic, individual differences in physicians' aptitudes—all contributing to errors. This chapter reveals the surprising clamor inherent in current human-mediated healthcare and why doctors struggle to muffle it.

A clinic cacophony

Patient care is riddled with inconsistency. In 2003 a landmark paper published in *The New England Journal of Medicine* concluded that patients in America only received recommended, evidence-based treatments about half the time.[2] In a follow-up study, in 2016, researchers at Harvard Medical School concluded that, "Despite more than a decade of efforts, the clinical quality of outpatient care delivered to American adults has not consistently improved. ... Deficits in care continue to pose serious hazards to the health of the American public."[3] Among their findings: around a quarter of patients failed to receive recommended cancer screening, 60 percent did not receive suitable treatment interventions for heart and lung conditions, and one in four Americans did not obtain adequate diabetes care. Research in other countries yields similar conclusions: physicians frequently fail to follow clinical guidelines, with non-compliance rates as high as 70 percent for some conditions.[4,5]

Variations between different doctors' judgments are another bass-note adding to the din in medicine. From cardiology scans to endometriosis tests, and from dementia diagnoses to psychotropic drug prescriptions, discrepancies exist between different doctors' judgments. For example, doctors' disagreements in radiology image reading have long been recognized as the "Achilles' heel" of the specialism. A major review of more than 12,000 imaging tests such as CTs and MRIs revealed that second reviewers disagreed with the initial assessments in one third of cases.[6] New interpretations changed the treatment plan in 18.6 percent of cases. Doctors even disagree with their own previous judgments. For example, a study of emergency physicians found considerable variability in how they later interpreted pneumonia chest images among children.[7]

Circadian rhythms, seasons, and even the time of admission can affect doctors' decisions. In primary care, the number of inappropriate antibiotics prescriptions written by doctors increases with each passing hour of the working day.[8] Or take another study of 50,000 patients across 33 primary care practices which found that referrals for breast and colorectal cancer screening decreased as the day progressed.[9]

Reflecting on all this noise, it is natural to wonder whether doctors should drink more caffeine and give themselves a stern pep talk, or whether some are plain inept. In this chapter we'll discover quite the opposite is true. Doctors' knowledge constitutes an outstanding feat of human achievement, but it is beset with mammoth challenges.

Striving to be in synch

What does it take to be an expert? Professor Anders Ericsson of Florida State University is one of the world's leading thinkers on human expertise—on how mediocrities metamorphize into maestros. On one level, Ericsson agrees with celebs like Lady Gaga: becoming an expert takes copious amounts of time—generally a decade or more—and heaps of hard work.[10] However, Ericsson says that true expertise demands more than just tenacious slog. To elevate oneself from middling ability to peak performance, he argues, requires a different kind of practice. Whether they recognize their technique or not, elites zealously push themselves out of their comfort zones in very specific ways. Distinguishing the Mozarts from the Salieris, top-flight experts work smarter by employing *purposeful practice*. This is marked by several key characteristics: well-defined goals, ongoing feedback, and focus.

Say as an adult learner you have mastered the Italian language with enough proficiency to get by. However, your goal is to speak like a native with perfectly rolled *rs*, an ear attuned to regional accents, and an impeccable facility with native hand gestures. The determined linguist won't just hone their skills with mindless repetition and wishful thinking. They practice in strategic ways to strengthen the weaker aspects of their performance until bad habits are ironed out, and good ones polished to become second nature.

To achieve seamless expertise, this goal must be broken down into smaller tasks. "Without feedback," Ericsson says, "either from yourself or from outside observers—you cannot figure out what you need to improve on."

Achieving "peak performance"—such as translating Italian so meticulously you could be hired by the FBI as a translator—demands never-ending self-critique. Experts do not rest on their laurels. Assiduousness about feedback facilitates a crucial quest to learn, finesse, and relentlessly improve upon their knowledge and skills. Even professional translators keep abreast of local idioms and dialects to avoid blunders.

Ericsson's insights shed light on what it takes to develop and sustain top-tier expertise across various fields. However, as a blueprint for what doctors should strive for, it falls short. Unlike music, sports, or foreign language learning, doctors face a much tougher mission to reach and stay at the top of their game.

In fact, the very idea doctors might aspire to be an all-star—a Lady Gaga or Michael Jordan of medicine—peddles a starry-eyed and deeply damaging ideal. The subtext is, with a little more elbow grease, a different style of working, or greater resolve doctors too can reach herculean heights of healthcare performance. The trouble is, from the first day they don their white coat till the day they hang up their stethoscope for the last time, the demands on doctors are supercharged.

Fledgling physicians and mixed ability learners

Student doctors swot up for years on anatomy, biochemistry, physiology, pharmacology, and pathology. Trainees are immersed in internal medicine, pediatrics, obstetrics and gynecology, psychiatry, general practice, and surgery. Clinical education is not just about committing biomedical facts to memory—doctors need to convert this into practical knowledge. Many medical schools now train fledgling physicians using high-tech medical mannequins programmed to show physiological reactions or even treatment responses. Students also cut their clinical teeth with the help of fake patients—actors who present with various ailments. These scenarios allow trainees to obtain much-needed feedback.

However, junior doctors cannot work with dummies or with fake patients forever. Like an adult gaining fluency in a new language, doctors must gain practice outside the classroom in the real world. Doing just that, each summer sees an influx of junior doctors to hospitals and clinics—a turnover that affects about 32,000 health workers in Europe, and more than 100,000 medical staff in the United States.[11] In the US, the annual rotation starts in July, giving rise to the phrase the "July effect"—coded clinical language for a substantial uptick in medical errors.

For some patients this yearly professional rite of passage will prove to be harmful or even deadly.[12] Fledgling physicians are associated with higher rates of misdiagnosis, spikes in medication errors, and increases in health-system inefficiencies. A study across five major American insurance companies of medical malpractice claims that closed between 1984 and 2004 concluded errors in judgment, lack of technical competence, and teamwork breakdowns were among the most common mistakes leading to litigation. The report found, "Lack of supervision and handoff problems were the most prevalent types of teamwork problems."[13] Both were more common among trainees.

No study (at the time of writing) has yet explored the impact of the July effect during the COVID-19 pandemic, which may have been worse than in previous years. Fast-tracked out of their degree program and working under savagely stressful conditions, it would be astonishing if a spike in errors did not arise. Or, as one doctor advised her X/Twitter followers, "Wear a mask ... unless you want to be intubated by an intern who did her last semester of med school via Zoom." Worse still, in countries where professional standards and training are not well established, or where there are fewer doctors to serve as teachers, error rates are likely to be even higher.

There is another bump on the road to expertise: junior doctors sometimes pick up bad habits. Doctors depict the life cycle of medical education with the slogan "see one, do one, teach one." This depiction overlooks a sticky problem associated with doctors' apprenticeships and their learning habits. Medical students inevitably emulate, gaze at, and nonconsciously imitate their seniors. The nature of this mimicry means that young doctors tend to copy both the positive and negative behaviors they witness—yet another source of noise.

Upon graduating from medical school, junior doctors spend between three and seven years in residency programs, acquiring skills in their chosen specialty. Although efforts are made to standardize education, another source of noise is the inevitable variety of training on offer. For example, Harvard Medical School, as many of its graduates eagerly assert, is an exceptional educational experience for a select few candidates. Access to this elite education benefits its graduates and, potentially, their patients, but its exclusivity provides little comfort to the broader patient community.

Or consider another kind of discrepancy—"board certification"—a voluntary process in the US whereby private medical organizations evaluate doctors' skills and award credentials in each specialty. This extra effort gives doctors a clinical edge, and those who mug up and pass these exams are entitled to call themselves "board certified." The extra credential is associated with improved adherence to evidence-based treatments and better patient outcomes.

Although it is not a legal requirement to practice, most US-based doctors—some estimates suggest around 90 percent—do obtain the qualification. However, it doesn't come cheap. Doctors strapped with higher levels of debt are less likely to be board certified across almost all specialties.[14] These doctors are also much more likely to care for minorities or patients with low incomes. Costs associated with board certification mean this system may contribute to health disparities. For example, a major study of more than 130,000 patients treated for heart attacks found Black Americans were more likely to be admitted to hospitals with lower rates of evidence-based treatments, and worse mortality rates than their white counterparts.[15]

Other factors create noise in care delivery. At a population level, female physicians may be more "fluent" in the language of medicine than males: on average, women are more likely to adhere to medical guidelines. Certainly, the suggestion here is not that every female doctor is superior to every male doctor. Knowing a physician's gender tells us nothing about their individual capabilities. Rather, pooling together the average performance and comparing differences at a population level, female physicians have a marginal—but not inconsequential—edge.

For example, one large-scale study followed the outcomes of over 1.5 million Medicare patients—the US national health insurance program for people aged 65 and over.[16] Investigators found that elderly patients treated by women were less likely to die within the

30-day period after hospital admission, and less likely to be readmitted compared with those treated by men. The difference in patient mortality was a modest 4 percent. Annually, however, more than 10 million Medicare patients are hospitalized—extrapolating from the data, if men achieved the same outcomes as women, 32,000 lives in America could be saved every year.

Forgetting words

After acquiring fluency in another language, as the speaker ages, they begin to forget words and phrases. The same is true of doctors. When it comes to mastering medicine, physicians suffer from a double dose of age-related troubles: the rookie problem, at one end of the scale, and the serious, yet taboo, challenge of cognitive decline, at the other. Biological aging is another source of noise.

Seniority is rightly viewed as a signifier of wisdom, but learned skills stagnate and expertise erodes. In an added twist of the scalpel, more mature doctors are not only more likely to commit medical errors but also more likely to express confidence in their judgments.[17]

In 2005, a review of forty years' worth of research found an inverse relationship between standards of clinical care and the number of years doctors had clocked on the job.[18] In 2018 a team of Harvard researchers explored doctors' competence with a random sample of over 700,000 US hospital admissions between 2011 to 2014, handled by nearly 19,000 physicians.[19] Among doctors aged under 40, the patient mortality rate was 10.8 percent—a figure that rose to 12.1 percent among MDs aged 60 or over. The numbers seem meagre, but the consequences are considerable. For every 77 patients admitted to the same hospital and treated by physicians aged 60 or older, we could expect one fewer patient to die if treated by doctors in their twenties or thirties.

People age differently, and doctors growing old need not in itself be a cause for concern. For example, when it comes to surgery, there is some evidence that doctors' skills don't erode as quickly.[20] In general, however, over time, the speed with which adults process information slows down and the capacity of our short-term and working memory diminishes. Even while there is considerable variability from person to person, cognitive functioning begins to decline in our forties.

Normal aging affects memory, vision, hearing, and motor skills. This means the capacity for continuous learning is compromised. On average, older doctors exhibit decreased professional performance across a range of clinical tasks, including gathering information from patients, keeping up to date with new biomedical knowledge, conducting physical examinations, record keeping, and diagnostic acumen.[21]

Growing old is inevitable and because of safety concerns in some high-stakes professions—such as air traffic control—the retirement age is legally mandated. However, in the ultimate high-stakes, life-or-death profession, awarding farewell carriage clocks is not compulsory. It's worth adding, however, that keeping doctors in their jobs as long as possible has been one way to cope with fewer frontline staff.

Still, across most countries there is no upper age limit for when doctors should hang up their stethoscopes. Quite the reverse: the trend is an aging medical workforce. In the US, since 1974, the number of practicing physicians aged 65 or over has increased nearly 100 percent.[22] In 2018, neurologist Dr. Gayatri Devi, an expert in memory loss, estimated that around 4,600 of the nearly 9,500 active doctors in the US aged 70 or over had Alzheimer's, and nearly 6,500 likely had some form of dementia.[23] Unfortunately, the lifetime risk of developing dementia in the US has only increased since these estimates were made, rendering patients even more vulnerable.

Disregarding grammar

Things get even noisier. Continuing with Ericsson's model of expertise, say a student has enrolled on an Italian language course with the goal of speaking like a native. The instructor offers a curiously pruned curriculum omitting vast chunks of teaching on nouns, tenses, and verbs. Naturally, by the end of the course we would not expect the budding linguist to emerge with a sound grasp of the language.

Something analogous happens in medical education. Doctors are taught from a structurally distorted base with implications for doctors' diagnostic checklists, treatment recommendations, and clinical judgments, and with harmful, even lethal, consequences for patients.

Gendered nouns

Consider a student attempting to master Italian by learning only male gendered nouns with scant emphasis on female nouns. This is what has happened through medicine's distinguished history. Clinical research and, consequently, doctors' education has failed to grasp that sex differences matter. One study found that images of biological male bodies were used three times as often as female ones in medical textbooks to illustrate "neutral" images of human anatomy.[24] Only in 2022 did medical students get to access a detailed 3-D female anatomy produced by leading medical publisher Elsevier.[25] Gaps in learning can directly lead to systemic sex-based blind spots in clinical practice.

Yet biological sex and hormones, even at a cellular level, can affect symptoms, immunity, response to infections, disease progression, and even how we respond to treatments.[26,27] Owing to biological sex differences people also experience different environmental and social stressors in their lifetime, as well as distinct vulnerabilities to disease. From auto-immune illnesses to depression, and from thyroid conditions to osteoporosis, the prevalence and risks of illness differ for women and men.[28] Nor can differences be safely ignored.

Lisa Marie Presley died of a heart attack in January 2023. Leading up to her death she reportedly experienced symptoms of heart failure that are unusual among men but more common among women.[29]

Men are more likely to experience crushing chest pains but women are more likely to experience what are, by the biological male default, "atypical" symptoms—neck, jaw, or upper back pain, shortness of breath, abdominal discomfort, nausea, or fatigue.[30] Onset of heart disease is also seven to ten years later than men. Although there is no evidence of misdiagnosis in Presley's case, because biologically male bodies are assumed to be the anatomical norm, women are more vulnerable to diagnostic and treatment delays. Even after hospitalization, women are two to three times more likely to die of a heart attack than men.[31]

Endometriosis is a debilitating condition where tissue similar to that which lines the uterus grows outside it, affecting other organs. It often causes chronic pain, heavy periods, and other symptoms, and can lead to infertility. The condition is believed to affect 10 percent of reproductive-age girls and women. In the US, patients wait an average of 8.6 years for diagnosis, with three quarters misdiagnosed.[32] In 2024, the UK Parliament's Women and Equalities Committee blamed "medical misogyny" for similar delays in diagnosing reproductive conditions.[33]

Even common ailments diverge. Pain is the most widely presented symptom in healthcare; women endure higher incidences of acute pain that last longer but female patients receive less pain-relieving medication.[34] However, women's pain is underestimated: doctors tend to believe they are "more likely to exaggerate" pain, which is also more likely to be discredited as "emotional."[35,36]

Clinical research also skews doctors' diet of learning. Throughout the history of medicine, most drug trials enrolled men, or used male animals, extrapolating findings to female bodies. In 1993, the Revitalization Act in the US attempted to address these baked-in biases by requiring representative inclusion of women, and racial and ethnic minorities, into federally funded clinical trials.

However, there is little evidence that these obligations are being met or successfully regulated by national research funding bodies.

For example, a study investigating high-risk medical device clinical trials conducted from 2010 to 2020 did not find an increase in female recruits: women made up only 33 percent of trial participants.[37] Under-recruitment of women in clinical trials means, treated as biologically male, female patients experience an almost twofold risk of adverse physiological drug reactions compared with men.[38,39]

Women bear the brunt of medicine's long legacy of one-sided curricula, but endemic educational biases can also affect men and boys. Eating disorders, for example, are often perceived as uniquely "female" illnesses, with the consequence that considerable numbers of male sufferers still go undiagnosed and undertreated.[40] While significantly more women contract breast cancer than men—1 in 8 compared with 1 in 800 men—this ratio does not place male diagnoses in the realm of rare diseases, of which more shortly. A US study in 2019 found that unique clinical characteristics of breast cancer in men, coupled with inadequate treatment, accounted for 63 percent of their higher mortality rate.[41]

Present, past, and future tenses

Medical education is refracted through a Peter Pan prism. Over-65s experience a higher burden of cancer, heart disease, and strokes than younger people. As we age, physiological changes necessitate treatments and drug dosages tailored specifically to older bodies. Yet an estimated 27 percent of the world's medical schools, and 15.8 percent of British medical schools, don't provide any teaching on geriatric medicine.[42] When they do so, it is often inadequate; one systematic review found that only seven medical schools worldwide implemented adequate education.[43] Despite aging populations, and fast-growing global need, the level of interest among medical students in geriatrics is low. In 2022–23 in the US, of 411 geriatric fellowships, 30 percent went unfilled.[44]

Over-65s, who make up the largest share of chronic illness patients, are often excluded or under-represented in clinical trials, despite accounting for 60 percent of new cancer cases.[45] An investigation into nearly 5,000 medical analyses of survival rates of patients with breast, prostate, and lung cancer from advanced phases of trials identified only a single paper, on breast cancer, where older patients were properly represented.[46]

Many older people have underlying health conditions and excluding them makes clinical trials easier to run and analyze. On the one hand, these exclusions are understandable; on the other, failure to recruit older people risks compromising the safety and usefulness of the results.

A review of medical research conducted between 2006 to 2014 recorded on the registry "ClinicalTrials.gov" found only 1.4 percent of the total number of 800,000 clinical trials focused on older patients.[47] An expert committee of the European Medicines Agency summed up the situation: "The drugs we are using in older people have not been properly evaluated."[48] Medications inadequately regulated for older age groups can cause "adverse drug events"—that is, serious harm or death. Conservative estimates suggest that 20–30 percent of patients aged 70–79 experience adverse drug events, with hospitalization rates due to harmful reactions four times higher than those of younger adults.[49]

Beyond sex and age, the inclusion of individuals from diverse social, geographical, and ancestral backgrounds is essential to determine the effectiveness and appropriateness of medical treatments.[50] Race is a complex construct with no categorical biological basis: what we consider "race" or "ethnicity" varies across history and geography, and is influenced by social and cultural factors. However, both historical and contemporary social influences undeniably play a causal role in shaping the demographic distribution of disease. Complicating matters, some inherited diseases are more prevalent among some populations—for example, sickle cell disease can occur among people of all backgrounds but is more common among people of African compared with European descent.

Although it is understandable why researchers and patients may wish to be "color-blind," important insights into the prevalence and treatability of illness are improved when medical research is mindful about diversity.

Despite this, in many countries, such as Canada, France, and across Scandinavia, national databases do not record race or ethnicity, preventing assessment of race-based health disparities. In the US, research into clinical trial participants shows recruits still fail to be representative of the wider population. In 2022, a report by the National Academy of Sciences, Engineering, and Medicine shone a light on the "critical shortcoming" of diversity in clinical research.[51]

Demographic biases embedded in clinical studies carry ramifications for the safety and usefulness of treatments that doctors are trained to offer. For example, pulse oximeters became crucial during the COVID-19 pandemic for detecting dangerously low oxygen levels, and a condition known as "silent hypoxia." However, these devices were not properly tested on darker skin tones. Due to inaccuracies in pulse oximetry readings, a study in the US found that Asian, Black, and Hispanic patients in intensive care units were more likely to receive less supplemental oxygen compared to white patients, potentially contributing to disparities in critical care outcomes.[52]

In the US, medical guidelines acknowledge illness disparities and adjust for vulnerabilities to disease. However, these risk scores often rely on outdated data or use race as a stand-in for more complex factors that can affect how diseases develop. For example, the American Heart Association's guidelines assign lower risk to Black patients without clear justification.[53] While direct evidence is limited, research suggests that such guidance may contribute to Black patients receiving less medical attention.[54]

Other evidence shows race-based medical myths, even if explicitly absent from curricula but not corrected by them, could influence doctors' decisions. In 2016, a study led by Kelly Hoffman at the University of Virginia found nearly three in four medical students and residents held at least one false physiological belief about African Americans, such as "Blacks' skin is thicker than whites'" or "Black people's blood coagulates more quickly than whites'."[55] Medical students who endorsed false beliefs were more likely to rate Black pain as lower than whites' and to make less accurate treatment recommendations.

Unique expressions

Worldwide, one in ten of us live with a rare disease—that's around 500 million people.[56] From Huntington's to sickle cell disease, and from Marfan syndrome to cystic fibrosis, many of us either live with a rare illness or know someone who is affected. In my own family, myotonic dystrophy type 1 ("DM1") affects my closest loved ones. DM1 is a serious degenerative, multi-organ condition that atrophies small muscles in the body.[57]

Globally, there are an estimated 7,000 rare diseases: around 350 comprise about 80 percent of the total.[58] This means many rare diseases are ultra-rare. There are also regional differences in rare disease classification. The EU categorizes a disease as "rare" if it affects no more than 1 in 2,000 people.[59] In the US a rare disease is defined as a condition affecting fewer than 200,000 people.[56] In Japan, a disease is considered rare if it affects fewer than 50,000 people—or roughly 1 in 2,500.[60]

People with rare diseases remain very much on the penumbra of medicine. A study in Australia found that 30 percent of people with rare illnesses waited between 5 to 30 years for a correct diagnosis.[61] In the US, a survey conducted from 2019 to 2020 by the National Organization for Rare Disorders reported that 50 percent of patients and caregivers attributed diagnostic delays to a lack of awareness about their disease, including among doctors.[62]

DIAGNOSIS AND TREATMENT

Gaps in diagnostic expertise are understandable because of the sheer volume of presentations and illnesses that physicians are required to master. Recognizing even a handful of rare diseases also depends on where the doctor is trained, their specialism, and the vicissitudes of personal interests.

Consider sickle cell disease again: it is common in some parts of sub-Saharan Africa, and in Nigeria the genetic illness affects around 1 in 50 births. In the US prevalence rises to 1 in 365 among Black Americans. Meanwhile in Europe, sickle cell disease is rare, affecting 2.6 in every 10,000 people. Cystic fibrosis, in contrast, is much more common in Europe and North America than in Asia or Africa; the highest rate of prevalence is in Ireland, where it affects around 1 in every 1,300 people. Adding to doctors' learning burdens, an estimated 250 new rare diseases are discovered annually.[63] In other words, the problem is getting worse for doctors, not better.

Deciphering even common maladies can be tough enough. So much so that doctors are explicitly taught to *avoid* entertaining the notion that patients present with rare illnesses. Recognizable to every medical student and physician is the maxim, "*When you hear hoofbeats, think of horses, not zebras.*" This quick-and-dirty heuristic is supposed to serve as a reminder that the most common causes are the most likely. It may serve most patients well. But it firmly relegates the already marginalized to second-class citizens in the clinic. Responding in kind, patients with the rare disease Ehlers-Danlos syndrome named their advocacy group "the Zebra Network."[64]

Beyond the often-welcome relief of a label, even when patients do receive an accurate diagnosis, doctors have very little to offer. Fewer than 1 in 10 rare diseases have a treatment or therapy associated with them. This is partly because pharmaceutical companies don't consider these populations as offering a lucrative return on investments. In the US, 95 percent of rare diseases have no FDA-approved treatment.[58] One review of clinical trials studying rare diseases found that more than half of studies were either not completed, or were left unpublished four years after the trial ended.[65]

Ever-evolving lingo

Aspiring to be fluent in Italian, a learner can be grateful about one crucial fact: the language will not change. Neologisms may emerge from time to time, but the syntax, phonetics, and lexicon remain relatively fixed and stable. Medicine isn't like this. Medical education is a bit like mastering late-medieval Florentine only to discover the lingua franca is now modern Italian. By the time medical students graduate, around 50 percent of what they've learned is out of date— though, as Canadian pioneer of evidence-based medicine David Sackett once counseled, they won't know which half.[66]

Recognition of the scale of the problem came about only recently. Stanford University Professor John Ioannidis set out to investigate the reliability of novel treatments, probing highly cited studies published in three top medical journals between 1990 and 2003.[67] He found that fewer than half of the original treatments were supported by further research, a quarter were never tested again, 16 percent were found to be less effective than first thought, and a further 16 percent were found to be completely ineffective.

Troubled by these findings, medical doctors Vinay Prasad and Adam Cifu delved deeper to gain some traction on the frequency with which treatments or interventions are later contradicted.[68] They sifted through all the original research papers published between 2001 to 2010 in *The New England Medical Journal*. In total, 363 papers tested an established medical practice; among them, 40 percent contradicted earlier published findings, with 38 percent confirming earlier studies.

Prasad and Cifu call these changes "medical reversals"—practices that were once considered unimpeachable and later discovered to be ineffective or even harmful. U-turns encompass medications, surgeries, and even public health programs.[68] For example, mammography among women in their forties may not improve breast cancer mortality but lead to higher rates of harmful and invasive treatment due to false positives—patients incorrectly diagnosed with cancer.

Just as mastering a language takes time, adapting to meaningful changes in medicine does too. U-turns in clinical treatments take almost a generation to become embedded. In 2011, a joint report by the Institute of Public Health at Cambridge University and the Rand Corporation estimated it takes an average of 17 years for clinical research to translate into guidelines and practice.[69]

Yet doctors face an ever-widening treadmill in their quest to become experts. Research publications are growing at an exponential rate, and primary care is unrivaled for the sheer volume of information to master. Very few studies examine how much doctors read and how well they keep up to date. In one US survey, conducted back in 1998, internists reported spending around four hours per week browsing biomedical articles, though 63 percent of them said they only read publication abstracts.[70] In another survey in 2002, 70 percent of US rehabilitation doctors said they "never read or scan" articles in medical journals, with 83 percent admitting they didn't read as much as they'd like.[71]

Across the whole of medicine, a new article is published every 39 seconds. At current publication rates—calculated using Medline indexing information—if doctors were to read just 2 percent of the present annual output of publications, this would require them to take 22.5 hours out of every day. This estimate is for scanning only a tiny percentage of the literature, leaving doctors a lavish one and a half hours each day to fit in work, rest, and play. It is therefore unsurprising that doctors tend to fall into the old, outdated medical vernacular they mastered early in their careers. Keeping up to date in any field of expertise is exacting; in healthcare it is insurmountable.

Lacking language coaches

Striving for fluency in Italian, the diligent learner needs dedicated teachers and conversation partners to point out mistakes. Recall that Ericsson's model emphasizes that both obtaining and sustaining expertise requires ongoing feedback. To achieve peak performance, doctors must also obtain first-rate coaching about how to improve their clinical decisions.

The problem is, doctors are not afforded ongoing moment-by-moment opportunities to identify, learn from mistakes, and hone their practice. This is yet another example of how medical expertise differs from other fields, creating more sources of noise.

After their apprenticeship finishes, doctors diagnose patients in health systems that are described as "open loop." Put another way: physicians obtain no systematic feedback on how well they are doing. As American humorist Don Herold once declared, "Doctors think a lot of patients are cured who have simply quit in disgust." Harvard Medical School patient safety expert Dr. Gordon Schiff explains:

> Typically, clinicians learn about their diagnostic successes in ad hoc ways (e.g., a knock on the door from a server with a malpractice subpoena; a medical resident learning, upon bumping into a surgical resident in the hallway that a patient he/she cared for has been readmitted; a radiologist accidentally stumbling upon an earlier chest X-ray of a patient with lung cancer and noticing that a nodule has been overlooked).[72]

Ericsson's model of expertise also explains why so much continuing education—the courses that doctors take to keep abreast of medical developments—is largely ineffectual. Traditional, passive forms of learning, such as attending conferences or workshops, offer scarce opportunities to improve clinical performance. Similarly, the custom of "medical rounds"—clinical lectures delivered by doctors—might benefit the résumé of the speaker but there is no evidence the ritual boosts physician learning or patient outcomes.[73,74]

Patient input can help but only goes so far. While many patients with chronic illnesses can become experts on their health, even they lack the luxury of the coach's perspective to pinpoint exactly what doctors could do better to fix mistakes or diagnose more accurately. Relying on patients also assumes they are forthcoming with feedback on errors, and that doctors are ready to listen.

Deprived of prompt, reliable, and efficient advice on their diagnostic accuracy, never mind pointers on how to improve on it, doctors miss a key element of top-flight expertise. Worse still, even if physicians were furnished with apposite, timely feedback on their performance, unlike golfers, musicians, or budding linguists, the volume of patients they see—combined with their schedules—renders it highly unlikely that beleaguered doctors could recalibrate and hone their expert judgments.

Unfortunately for us, the challenges do not end here. Even if medicine could miraculously create a Dr. Amazeballs of medical expertise, and avoid all this noise, another problem awaits: how doctors think. This is the focus of the next chapter, where we'll discover that some of us are treated worse than others.

7

The Dark Art of Medicine

In tennis rankings, Serena Williams is one of the all-time greats. Champion of 23 Grand Slams, she holds the record for winning the most singles titles of any female player in history. In 2017 Williams contested the most crucial line call of her life—this time, off court.

A day after giving birth to a healthy baby girl, and still resting in hospital, the tennis legend experienced some ominous but familiar symptoms—tightness in her chest, and shortness of breath. Because of her C-section surgery, Serena was off her regular daily dose of anticoagulants but with a history of pulmonary embolism—or blood clots in her lungs—she immediately understood what the symptoms meant. Promptly reporting her concerns to clinicians, Ms. Williams' worries were abruptly dismissed. Not taking "no" for an answer, the tennis champ's renowned tenacity kicked in—"*I told you*, I need a CT scan and a heparin drip."[1] Confirming her first instinct, doctors identified and treated potentially lethal blood clots in Serena Williams' lungs.

Keris Myrick is a self-confessed "army brat." Born in Bremerhaven, Germany in 1961, because of her father's military service, the Myricks moved all over the world. Now based in Los Angeles, Keris still travels widely, giving talks on patient advocacy. She sits on the boards of the Center for Health Care Strategies and Mental Health America, and is Vice President of the board of Disability Rights, California. Keris speaks from experience. Forty years ago, aged 21, she was diagnosed with schizophrenia and today she lives with diagnoses of schizoaffective and obsessive-compulsive disorders. Keris also has experience of physical ill health. Nearly a decade after first experiencing unusual symptoms in her throat, she was diagnosed with thyroid cancer.

"I was complaining about this thyroid issue for years," Keris tells me, "but doctors said it did not exist. And I kept saying, 'There's a lump in my throat when I swallow. It gets stuck at the top, and I have to kind of manipulate to the left or to the right.' I said, 'It feels like it's a crooked tunnel.'"

"I was choking on food ... I had to go to the hospital three times to have food removed from my esophagus."

"But at every point, people kept seeing ..." Not one to mince her words, Keris interrupts herself, "Quite frankly, this is my fear: when they see you have a mental health condition, particularly schizophrenia, then you're filtered through that diagnosis: 'Ah, yeah, you have schizophrenia. That's paranoid thinking. Oh no, that's a delusion. Oh, that's not real.'"

"And it turned out the cancer was absolutely real."

Dr. Keith Geraghty is from Dublin, Ireland and works as a Research Fellow at the University of Manchester where he is an expert in myalgic encephalomyelitis or "ME", sometimes referred to as "chronic fatigue syndrome." ME is not fully understood but in many instances it is caused by viral infections, with similarities shared with Long COVID.

Studies also show patients who develop these conditions have underlying immune, genetic, or physiological vulnerabilities.[2] Bedbound and housebound, many patients live with chronic pain and with little energy to devote to even the most mundane and basic of everyday tasks.

Back in 2004, Keith was enrolled in medical school, striving for the job he had always wanted. Not only academically gifted, he also excelled at sports: Keith was the captain of his field hockey team, a university rower, and an army reserves recruit. But his studies were devastatingly halted by ill health. In short, Keith's interest in ME is not just academic, it is personal.

Despite his physical fitness, when he was first diagnosed his primary care doctors wrote a dismissive report describing him as "an overachiever type." They claimed he was using his illness as an excuse not to keep up his own high standards. Keith was immediately referred to a psychiatrist for treatment. He told me, "Doctors insisted the illness was all in my head."

It seems natural to wonder: Were Serena Williams' concerns dismissed as mere effects of maternal hormones? Could Serena and Keris—who is also Black—have been victims of racial profiling? Did Keris's mental health problems overshadow her cancer diagnosis? Was there something about Keith's symptoms that led doctors to discredit his genuine illness?

Abundant evidence shows that our appearance, and even our symptoms, can influence the quality of care we receive—including whether we receive medical attention at all. However, establishing prejudice in healthcare isn't always straightforward.[3] As we'll discover, first-person perspectives offer a crucial starting point, but even these observations don't always reveal the subtle nuances of discrimination. Fortunately, science can help iron out the ambiguities that wrinkle individual anecdotes, even while blots of uncertainty remain.

Off-color visits

Consider what happens before we meet a provider. Like the Roman Emperor Nero, who with the meagre gesture of a thumb-up or thumb-down determined the fate of gladiators, a variety of gatekeepers can influence patient destiny. In a study conducted in Toronto, Canada, researchers placed phone calls at 375 general practices requesting to see a primary care doctor.[4] Confederate callers followed a script designed to indicate whether they were from high- or low-income brackets. To suggest wealth, they greeted the receptionist by saying, "I was just transferred to Toronto with [name of major bank], and I need a family doctor." In contrast, low-income callers said, "I'm calling 'cause my welfare worker told me that I need a family doctor." All would-be patients inquired, "Is Dr. [name of doctor] accepting new patients?"

Despite Canada's universal-access single-payer healthcare system, and without uttering a word about their health status, "wealthier" callers were 50 percent more likely to be given a medical appointment than their "working-class" peers.

Might patients fare better when clinicians man the phones? Sociologist Heather Kugelmass examined whether class, race, or gender might influence access to mental healthcare.[5] Enlisting actors to pose as patients requesting appointments in New York City, she asked confederates to adopt specific race- and class-based speech patterns, and to read scripts with racially distinctive names. To ensure that the manipulation was successful, and voices seemed authentic and representative of the target demographics, Kugelmass used a crowd-sourcing site to pool feedback from the public. To guarantee that appointment decisions were made by clinicians and not office staff, calls were restricted to those who ran solo practices and placed with more than 300 accredited psychotherapists with a PhD or PsyD degree in New York City.

Kugelmass found that callers' gender did not affect decisions, but a patient's class and race did. "Middle-class" patients received appointments nearly three times more often than their "working-class" counterparts; middle-class "Black" patients were also less likely to be offered a mental health appointment than their "white" peers.

Even fresh-faced medical students may succumb to clinic classism. In a study in Canada, trainees were randomly assigned to watch a video of a doctor talking with either a "wealthy" or a "working-class" patient whose socio-economic status was manipulated through clothing, accessories, and a carefully scripted dialogue matched for the same medical complaint—chest pain.[6] Only on the cusp of their clinical vocation, students allocated to the working-class patient were less likely to agree that "This is the kind of person I would like to see in my practice."

Do classist attitudes arise during our appointments? An interview study with Medicaid patients—the US federal and state medical care for patients on low incomes—found many patients perceived that physicians treated them differently.[7] Perceptions were represented by a wide array of derogatory terms, painting a vivid and disparaging picture: some felt they were "on the back burner" or viewed as "a bum," others felt devalued as a "leech," "livestock," "a number on a file," "peasant," or "scum." Other studies show that relative to patients with high incomes, those with lower incomes are more likely to perceive prejudicial treatment.[8,9]

Communication is a two-way process, and muddying matters, some patients are more passive than others. Audio-recorded studies show that those with lower incomes may be more reticent during appointments than their richer peers. A survey in Britain found that people with lower incomes were more likely to feel embarrassed and less confident when discussing symptoms associated with cancer.[10] Or take another study of patients with breast cancer, which found that those with a low income or fewer years of formal education asked fewer questions.[11] Doctors tended to reciprocate by providing an average 153 utterances of information with lower-income patients compared with 228 among richer patients. Better-educated patients also fared better, receiving an average of 207 nuggets of health information compared with 165 among those with fewer years of formal education.

In the US, Black women face the most severe maternal mortality rates among all racial and ethnic groups. Researchers point to higher rates of pre-existing health conditions and the enduring effects of systemic racism as key contributors.[12,13] A recent meta-analysis found that patients who reported experiencing racial discrimination in society were more likely to have poor pregnancy outcomes.[14]

In the considerable press coverage following Serena Williams' neonatal experiences, many journalists wondered if her treatment by health professionals was also racist. Patients in the US who identify as Black or African American are more likely to detect subpar treatment by doctors.[15] In 2020, a Pew Research Center survey found that 75 percent of white Americans reported a "mostly positive view of medical doctors," dropping to 68 percent among Hispanics, and 61 percent among Blacks.[16]

Analyses of electronic health records may provide support for patients' perceptions. Controlling for sociodemographic and health characteristics, a large study led by researchers at the University of Chicago analyzed 410,113 clinical notes about nearly 18,500 patients. In their investigation they controlled for factors such as patient sex, insurance type, and health status. Even after taking these steps, compared to documentation written about white patients, clinical notes about Black patients were 2.5 times more likely to include negative descriptors such as "noncompliant" and "unpleasant."[17]

Even in countries with universal healthcare, people report being treated differently. In a study of nearly 140,000 cancer patients across 144 hospital trusts in England, ethnic minorities reported lower satisfaction with their care.[18] White British or white Irish patients were the most satisfied while Black African, Black Caribbean, and Asian including Chinese, Indian, Bangladeshi, and Pakistani patients were less than half as likely to describe their cancer care as "excellent."

We cannot be certain about why Serena Williams was not initially listened to. However, in 2019, a US study found Indigenous, Hispanic, and Black women experienced higher rates of racial mistreatment in maternity care—such as being scolded, being threatened, or receiving no response to requests for help.[19] Mistreatment was more common among patients with lower incomes, but this didn't explain all of the differences. Similarly, in the UK in 2022, a major study of maternity care found that non-white patients were more likely to report being ignored and disbelieved, with many denied pain relief.[20]

Patient race and ethnicity may also be associated with higher rates of medical errors.[21-24] In the US, one study analyzed hospital data collected between 2007 and 2011 to explore possible racial differences in opioid prescriptions in emergency departments.[25] After controlling for age, sex, health insurance, year of treatment, and hospital location, investigators did not find any race-based differences in pain administration when patients presented with "objective" conditions—such as a broken bone or kidney stones. However, when presenting with less easily identifiable complaints—for example, back or abdominal pain—Black patients with the same level of discomfort were half as likely to receive pain medications as whites.

Racial concordance—sharing the same race or ethnicity as your doctor—might play a role.[26,27] A study in Oakland, California, found that Black men paired with Black doctors were more likely to choose preventive services like diabetes screenings and flu shots.[28] Researchers estimated this could reduce the Black–white male cardiovascular mortality gap by 19 percent.

In 2018 a systematic review of 40 studies concluded that racial concordance was "clearly associated with better communication."[29] However, some researchers argue that racial concordance studies often confuse correlation with causation, with other factors driving these differences.[30]

The contours of the visit are undoubtedly complex. For example, Howard Gordon from the University of Illinois analyzed conversations between lung cancer patients and their doctors.[31] Black patients were provided with less medical information compared with white patients; however, these patients were also less proactive in seeking advice. The reasons for these communication differences are unclear. For example, it's possible some patients from racial or ethnic minorities feel heightened anxiety during appointments, which may limit their participation.

Compounding matters, misperceiving passivity as disinterest might prompt a detrimental downward spiral. In a carefully constructed study, Richard Street and colleagues studied more than 200 interactions drawn from visits at 10 outpatient clinics and concluded that "reciprocity and mutual influence" between patients and doctors influenced the ebb and flow of dialogue.[32] Doctors assumed patient-friendly communication with those they considered more actively engaged or better communicators, or those they believed would better adhere to treatments, and were "more contentious with contentious patients."

The bottom line is, patients who are already marginalized during visits can fare the worst.

Clinic caste systems

Some conditions are just medically "uncool." That might sound like an inflammatory comment, yet surveys conducted by sociologist Dag Album in egalitarian Norway expose what most doctors and many patients have long intuited. In three rounds of questionnaires—conducted in 1990, 2002, and 2014—doctors were requested to rank diseases according to "the prestige you imagine it has among health personnel."[33,34] Album found scorecards across the decades remained "clearly and consistently" stable. The medal-winning maladies—considered the most prestigious diseases—were leukemia, brain tumors, and heart attacks. Medical booby prizes went to fibromyalgia, depression, anxiety, and cirrhosis of the liver.

We might suppose ratings reflect deadlier diseases. For example, heart disease tops the prestige list and is one of the leading global killers, causing one in four deaths annually in the US. Yet, challenging this assumption, bottom-tier illnesses depression and cirrhosis account for over twice as many deaths as leukemia and brain tumors combined.

According to Album and his colleagues, illness rankings tell us more about doctors and the practice of medicine than they do about patients and their problems. Disease classifications, he claims, are instead associated with doctors' prestige, with starring roles reserved for the "acuteness and drama" of some diseases while lower-caste conditions are considered "chronic and mundane."[33]

Whether this critique sounds like a fair appraisal, or an overwritten theatre review, there is no question that, for doctors, not all illnesses, and therefore not all patients, are equal. Dr. Keith Geraghty, the Irish researcher, experienced feelings of ridicule and time-wasting when he visited his doctor with ME. Willing to prove he had nothing to hide, Keith attended sessions with a psychiatrist where he was told not to think he had a physical illness, and to keep a diary of his health. "None of this helped," Keith told me. "In fact, I kept getting worse, until such time as I was bedbound most of the day."

Keith isn't alone in this. Survey research shows people with chronic pain conditions and medically unexplained conditions such as ME and Long COVID are particularly vulnerable to negative labeling, with their complaints commonly attributed to their poor mental health.[35] In Britain, a recent survey found that most patients with ME were dissatisfied with the care they received.[36] While studies reveal that many doctors would "do anything for these patients," substantial numbers of clinicians describe people with ME as *"heartsinky"* and a "burden," with patients' character—as suspected malingerers or lazy freeloaders—often called into question.[37]

Beyond ME, doctors can be as judgmental and stigmatizing as the general public, including about mental health.[38,39] Do these attitudes bleed into practice? Patient champion Keris Myrick suspected that her thyroid complaints were overshadowed by her schizoaffective diagnosis. Negative perceptions about patients with psychiatric conditions and misattributing symptoms to mental health conditions can contribute to errors.[40] For example, compared with other patients, those presenting with serious mental illness and diabetes at emergency departments are less likely to be admitted to hospital for diabetic complications.[41]

Again, analyses of electronic medical records reveal that negative descriptors (e.g., "noncompliant," "uncooperative," "nonadherent") are more commonly applied to patients with diabetes, substance abuse disorders, or chronic pain.[42] This language might be associated with inferior care. For example, in a US survey of 655 emergency medicine doctors, those who used the term *"sickler"* to denote people with sickle cell disease were more likely to express negative attitudes about patients and to adopt less intensive treatment recommendations.[43]

Patients with disabilities are also at higher risk of injury through neglect, delayed diagnoses, or failure to provide adequate treatment, including preventive care. One study found that after being diagnosed with early stage lung cancer, people with physical disabilities were significantly less likely to be offered life-saving surgery than able-bodied patients.[44] Female patients with physical disabilities are 70 percent less likely to be offered contraceptive pills than other patients, less likely to undergo mammography or cervical "Pap" smear screening,[45] and following early stage breast cancer, are less likely to be offered breast-conserving surgery compared with able-bodied patients.[46]

At the start of the pandemic, the COVID-19 Disability Rights Monitor Dashboard received complaints from people in 134 countries worldwide relating to their treatment, and by October 2020 its report highlighted the "catastrophic global failure to protect the rights of persons with disabilities,"[47] findings that appeared to bypass major news outlets. In some regions, clinical triage decisions deprioritized people with various categories of disability,[48] even considering some patients ineligible for ventilators.[49] The soft bigotry of low expectations can creep into clinicians' decisions too. A US study reported that patients with mobility problems were 20 percent less likely than other smokers to be asked about their habit during health check-ups.[50]

In 2021 Dr. Lisa Iezzoni led a national survey of doctors' perceptions of people with disability, and only around half of those surveyed "strongly" agreed that they would welcome patients with disability into their practice.[51] Typifying the response, one rheumatologist admitted that when they see a patient with a disability on their daily schedule, "They groan or are not as excited … getting through the visit is going to take longer."[52]

Age, weight, and sex talk

The term "ageist" is relatively modern, first coined in 1969 during an interview between American gerontologist Robert Butler and then cub reporter Carl Bernstein at *The Washington Post*.[53] Nearly two decades later, a pioneering study into elder discrimination used videotaped visits. Researchers reported that doctors were "less respectful, less patient, less engaged, and less egalitarian with their old than with their young patients."[54]

Since then, several studies have identified "elderspeak"—a distinctive sing-song parlance characterized by louder, slower speech and baby-talk tones—often directed at older patients by health providers.[55,56] In defense of clinicians, slower repetition of information by caregivers is shown to improve recall,[57] but condescending cadences, research shows, rankle retirees. Suffering this indignity in silence, surveys show goo-goo-ga-ga-ing doctors are not only viewed as disrespectful, but are also considered less competent than unaffected clinicians.[58]

Establishing the causes of health disparities is never easy but there is evidence that ageism affects treatment decisions. For example, in the UK, older people with lung cancer are less likely to be referred for surgery, despite evidence that such procedures carry similar success rates as for younger patients.[59] While health outcomes following a stroke are not age-dependent, a Europe-wide investigation revealed seniors were offered fewer rehabilitation and treatment services than their younger counterparts.[60]

Physical appearance affects doctors' opinions about us in other ways. Worldwide, more than 1 billion are now considered medically obese, and in the US and Europe most adults are now overweight. The causes seem obvious: overeating and lack of exercise. However, common sense doesn't always cut nature at its joints, and evolutionary psychology challenges the simplicity of conventional wisdom, with scientific accounts going beyond weak willpower. Despite pop-psychological theories, stigmatizing obesity is not a useful strategy to tackle it, with some evidence showing it is counterproductive.[61]

When it comes to weight gain, epidemiology shows that genetics plays a significant role, as do social and environmental factors such as sleep deprivation, shift work, the cost of healthy food, social status, and living in pedestrian-friendly neighborhoods.[62–64] Science, however, has not softened the court of public opinion. Research shows that obesity tends to elicit feelings of disgust and contempt, and perceptions of lowly character.[65]

People carrying extra pounds often look to doctors for help, but healthcare isn't always a safe haven either. Cross-cultural studies among primary care physicians show anti-fat sentiments are far from taboo.[66,67] In one US study, as patients' body mass index increased, physicians' desire to help decreased.[68] Another survey of primary care doctors found that more than 50 percent viewed heavier patients as "awkward," "noncompliant," and "ugly," with around one third viewing such persons as "lazy" and "sloppy."[69] Or consider a study of nearly 5,000 medical students drawn from 49 American medical schools: two thirds explicitly embraced discriminatory views, with one in six medical students agreeing with the statement "I really don't like fat people very much."[70]

Every two-faced gossip knows it is one thing to harbor negative opinions but quite another to voice them. Do these attitudes influence visits? A British study found that patients were deeply reluctant to discuss weight loss with their doctor because medical advice tended to be negative or unhelpful; as one interviewee recalled, "It was more like, '*You've got a big problem. Sort it.*'"[71] Another videotaped study of primary care appointments found doctors spent significantly less time educating heavier patients about their health.[72]

Sex affects talk and treatment too. In the West, where most research into patient–doctor communication is conducted, men and women report comparable levels of satisfaction with their care.[73,74] However, patients who share the same gender as their doctor may experience a moderate boost. One review found that same-gender pairings between patients and doctors tended to be characterized by more relaxed and longer consultations.[75] Female doctors appeared especially well tuned to women's needs, values, and preferences. Meanwhile, male-to-male medical pairings prompted more discussion about patients' social problems.

When we delve into diagnosis and treatment, problems emerge. A major study of over 580,000 heart attack patients admitted to Florida hospitals from 1991 to 2010 found that female patients had a higher mortality rate when treated by male doctors.[76] In total, 13.3 percent of women died compared to 12.6 percent of men when treated by male physicians. There was a smaller gap when female doctors were in charge: 11.8 percent of men died, compared with 12 percent of women.

Just as Keris experienced, and possibly Serena Williams too, "medical gaslighting" refers to those occasions when presenting symptoms are wrongly attributed to emotionality, hysteria, or "baby brain" rather than to underlying health problems.[77] In 2019, asked whether they felt they had ever been ignored or dismissed, or felt like they had to prove their symptoms to their doctor, one survey found 17 percent of women agreed compared with 6 percent of men.[78] Another study in the US found female patients were more likely to have their symptoms reported in their medical records with sarcastic quotation marks.[79]

In the West, negative attitudes toward sexual minorities have declined considerably;[80] but in medicine, discrimination may still arise. In 2017, a survey in the US found that 8 percent of lesbian, gay, or bisexual respondents were refused care by a doctor because of their sexual orientation; one in five people identifying as transgender reported that clinicians directed harsh or abusive language at them during the visit.[81]

In summary: first impressions have the power to influence how the visit unfolds. Indeed, attractiveness biases in medicine have not been systematically explored and it would be surprising if these also didn't affect treatment, given that beauty positively influences employment and other high-stakes decisions.[82] Accent has not been studied in relation to clinician biases and patient outcomes, but studies show it can be a potent cue for discrimination in other settings.[83]

To understand why these discrepancies arise we need to closely examine how doctors think. Before we delve into this, let's first examine how doctors *think* they think, which offers a window into why things go wrong.

Doctors musing on their "art"

Interviewed about his artistic process, Mick Jagger once explained, "The bottom line is: songwriting is songwriting. ... You write with different people, you get a different flavor."[84] Famously terse in interviews, when probed on the matter, musician Van Morrison was more direct, though equally illuminating: "I can't explain it. ... See, I don't need to explain what I do or how I do it, I don't need to."[85] Frank Zappa once opined, "Most rock journalism is people who can't write, interviewing people who can't talk, for people who can't read." More charitably, musicians, artists, and other talents don't owe us psychological portraits of their creative processes. All that matters is performance.

Responding to a similar question, doctors are no more edifying than rock stars. In a vox pop study at a university in Canada, 50 physicians, of all ages and seniorities, were asked what they understood by "the art of medicine." You can view their wide-ranging responses on YouTube.[86]

Like popstars, doctors—from students to medical academics—offered confident if inchoate commentary: "The art of medicine," expounded one, "is synonymous with its originality, imagination, inspiration, innovation, and ingenuity." Another proclaimed, "It's not being afraid to go outside of what is routine and to be colorful. ... The ability to be artistic is very important." One professor of medicine took the minimalist and succinct Mick Jagger route: "The art of medicine is about communication." Said another: "It's about developing your own style." Channeling Van Morrison, one academic doctor said: "It's hard to put your finger on but you know it when you see it."

On the strength of these interviews, we might be forgiven for thinking—to borrow Marshall McLuhan's phrase—that the art of medicine is "anything doctors can get away with." Unfortunately, there is a sense in which this holds true. Talented people can be frustratingly poor at articulating their expertise—whether it is songwriting or clinical judgment. More generally, humans are opaque and fuzzy when furnishing explanations about their thought processes. Although we have privileged access to how we feel and what we think, we don't have reliable, introspective access to all the reasons for and causes of our innermost thoughts, feelings, or hunches. In short, doctors, much like pop musicians, skilled performers, and many others, often struggle to track, trace, and deeply explain the reasoning behind their actions.

This vagueness can mask a multitude of medical sins. Moreover, as cognitive linguists point out, the very language we use can offer a vista into how we think.[87] When doctors use the metaphor "art" to describe their expertise, it comes with conceptual baggage, conjuring up ideas about "talent," "creative flair," and even idiosyncratic styles of practice. Conceiving their expertise as an "art"—something Delphic and enigmatic—doctors unwittingly celebrate, exalt, or even glorify the very real "black-box" challenges that lie at the heart of what they do. So, let us shine a light on medicine's dark art.

DIAGNOSIS AND TREATMENT

Medicine as a dark art

Research identifies around fifty covert, unwanted biases that interfere with doctors' judgments during our visits. For example, premature closure is the tendency to reach a clinical decision before gathering all the necessary information or exploring other possibilities. Availability biases arise when doctors make a diagnosis simply because it is retrieved from memory faster. For example, prior to reliable tests for COVID-19 it was easy to diagnose every patient with shortness of breath, dry cough, and a rash as having contracted the illness. During the pandemic doctors sometimes misdiagnosed the virus, confusing it with conditions ranging from lung cancer to acid reflux and from severe allergies to asthma.

Confirmation biases, in contrast, arise when doctors selectively cherry-pick information to reach a diagnosis. Presented with complex clinical scenarios, they tend to prop up their first hunch, dismissing information presented later in the process.[88] A patient with constipation and acute abdominal cramps, with a family history of bowel complaints, may lead doctors to discount other possibilities, overlooking appendicitis.

Doctors are trained professionals, so why do they succumb to such troublesome quick-and-dirty biases? Turning the tables, doctors' decision-making depends on quick-fire judgments. All humans rely on biases and heuristics.

A quick glance back in time is necessary. The capacity to stereotype and quick decision-making were crucial for survival. Based on a limited set of observed characteristics, humans made—and continue to make—practical inferences about situations we have not observed—a skill that was, and still is, vitally important for getting by. Take the most wholesome example: picking an apple to eat. How did our ancestors know what was apple-y?[89] They relied on stereotypes: the round shape, the smooth texture, the sweet taste. Generalizing from a few characteristic features reliably allowed them

THE DARK ART OF MEDICINE

to identify apples and grasp that they were edible. Stereotypes both enable and constrain our perceptions of the world, and it is this aptitude that underpins our ability to distinguish apples from rhubarb, rabbits, rocks, rivers, and all manner of other stuff.

Though it's un-PC to say, we also categorize people: the Irish are friendly; Italians are passionate; Americans are loud; and the British talk about the weather. Stereotypes range from the bland to the celebrated and can be toyed with or shoved on for effect. Social stereotypes can be crass, or plain wrong. In the media, Black Americans are portrayed as more likely to abuse illicit drugs than their white neighbors yet rates of substance abuse are comparable.[90] Americans might imagine sub-Saharan African immigrants to be uneducated but they are among the most academically accomplished people in the US.[91] The Irish are renowned for merrymaking yet, with high rates of depression and suicide in Northern Ireland, not everyone is "having the craic."[92]

Even while it may affront our sensibilities, stereotyping has also been subjected to, well, *stereotyping* and cannot be easily pigeonholed. Like a smudged mirror, some stereotypes capture statistical data fairly well.[93] In the US, for example, there are genuine social differences along the lines of race and ethnicity. Black Americans really are more likely to be incarcerated than whites.[90] Asian Americans academically outperform other racial groups.[94]

Owing to socio-economic societal differences along demographic lines, it would be surprising if there weren't increased disease susceptibility for some people. In addition, just because some stereotypes are unflattering or troubling does not make them false. Women really are twice as likely to suffer from depression than men.[95] In the US, after controlling for income and health insurance coverage, Black and Hispanic Americans are considerably less likely to take their meds than white Americans.[96] Patients with severe mental illness really are less likely to take the psychotropic medications their doctors prescribe.[97]

DIAGNOSIS AND TREATMENT

Again, the mere presence of patterns of behavior or of stereotypes doesn't explain why they exist. Take one source of stereotyping—"coalitional psychology"—our instinctive tendency to cozy up to some and recoil from others. For most of our 200,000 years on Earth, humans lived in small, semi-nomadic bands where our ancestors knew or recognized most of their neighbors. As a legacy, even today, we systematically *over*value the performance, qualities, and abilities of people we judge as like us. Captured by the proverb "birds of a feather flock together," this phenomenon—termed "homophily" by social psychologists—refers to our predilection to trust and respond favorably to people who resemble us.[98,99] This old psychology can be activated by accents, language, clothing, political affiliations, and religious insignia.[100] In fact these tendencies are embarrassingly easy to elicit in all manner of social fora, including football matches, job interviews, social media, or in university lab experiments populated by ostensibly progressive students.[101]

From antiquity to the present day, stereotyping has influenced how societies are constructed—often with tragic social consequences of degradation for some groups of people. However, biology is far from destiny, as evolutionary psychologists are hoarse from pointing out.[102,103] Consider racism: there is nothing inevitable about racial bigotry from an evolutionary perspective.[104,105] It is highly doubtful that our ancestors met "racially" different people, and there is no evidence that natural selection adapted our mindware specifically to code other people's "race" or "ethnicity."

Doctors, however, are only human, and they engage in stereotyping too. This is further complicated by the fact that their very expertise depends on it. The same human capacity to pigeonhole is what powers our doctors' diagnostic acumen. Differentiating when biases are reasonable and when they are problematic is a massive challenge for clinicians.

Insights from the late Nobel Laureate, cognitive scientist, and author of *Thinking, Fast and Slow* Daniel Kahneman illuminate why.[106] According to Kahneman the human mind comprises two systems of thinking—hence, "thinking fast" and "thinking slow." These systems are a bit like the fabled hare and tortoise of Aesop's tales. Fast thinking (the "hare")—also referred to as "System 1"—uses rapid rules of thumb to make nonconscious, automatic, and uncontrolled decisions. Slow—"System 2"— thinking, on the other hand, is the tortoise of mental processing. Verbal, conscious, and deliberative, this system takes time and effort.

Doctors' expertise constitutes a form of subtle pattern recognition: they explicitly learn to associate clusters of symptoms and signs with specific illnesses, and illnesses with specific treatment plans. In the previous chapter we discovered that acquiring expertise is highly demanding, requiring committed practice and deliberate, rigorous learning. Eventually, over time, this investment leads to the emergence of intuitive perceptual patterns, enabling doctors to "read" symptoms automatically and nonconsciously. Emory University cognitive scientist Robert McCauley refers to this as "practiced naturalness."[107] In the terminology of Kahneman, what begins as the slow, deliberative learning of facts—System 2 thinking—gradually transforms into rapid, instinctive, life-saving System 1 responses.

Where does this leave "the art of medicine"? From a psychological standpoint, medical knowledge is System 1 thinking—knowing without knowing—where decisions are made swiftly and nonconsciously. This fast thinking is indispensable to doctors' jobs. If they were to deliberate over every minute detail of patients' presentations via some kind of inner Socratic dialogue, doctors would lose the ability to make quick decisions under pressure.

Mental shortcuts come with a cognitive catch. System 1 is a black box. Doctors know what they know, but not how they know it. Stereotyping, as we earlier noted, carves up the world but it doesn't come with guaranteed precision, nor is it tagged with precise interpretations about why its categories arose. Because doctors depend on fast thinking, they tend to be unaware about how unwanted biases can insidiously tamper with their judgments.

In other words: System 1 invites a dilemma. The challenge is for clinicians to sort the cognitive wheat—*good biases*, from the chaff—*bad biases*. Can System 2 come to the rescue? Analogies are never perfect. Unlike the tale of the hare and the tortoise, human cognition is not a subcranial competition of "hasty" System 1 racing the "wiser" System 2. The latter is not a steady, high-voltage light that can illuminate the inner black box of our mind, showing System 1's workings. Instead, it gives us a convincing but after-the-fact narrative about System 1—one that feels totally plausible but doesn't actually reflect what happened.

If System 1 is an impetuous buyer, System 2 is a salesman in an expensive suit; as Kahneman writes, System 2 is an apologist for System 1 rather than an astute critic of it—"an endorser rather than an enforcer."[106] Similarly, doctors might explain their decisions after the fact, but this doesn't mean they've given a truthful account of their thought processes.

Broad brushstrokes

Can anything be done to remove the unwanted biases in doctors' thinking? Members of the medical community have rolled up their starched white sleeves and proposed some remedies. In the last decade, "anti-bias" training classes have spread rapidly, including in medical schools. A key idea behind this approach is that doctors can be taught to identify their own hidden prejudices. According to conventional wisdom, with better awareness, doctors will be in a stronger position to override unwanted stereotypes. For example, the Ethics Code of the American Medical Association (AMA) advises doctors to "Examine their own practices to ensure that inappropriate considerations . . . do not affect clinical judgment" and more plainly to "Avoid stereotyping patients."[108]

As we've seen, doctors cannot usefully avoid stereotyping, otherwise they'd cease to do their job. The AMA's advice therefore restates the problem. Besides, System 1 thinking operates below the level of conscious awareness, so self-inspection is an unproductive tool for excavating problematic stereotyping. Various attempts to thrust nonconscious, System 1 thinking out of its black box onto the well-lit arena of conscious awareness often fail to acknowledge, as Kahneman emphasizes, that "Biases cannot always be avoided, because System 2 may have no clue about the error."

Worse still, there is a lack of evidence on the long-term effectiveness of anti-bias training, an issue that is rarely addressed in medical journals or curricula.[109] Such training might even be harmful.

Musa al-Gharbi, a sociologist at Columbia University, has extensively researched the literature on anti-bias training and argues that promoting awareness about the prevalence of stereotypical beliefs could be counterproductive.[110] Drawing on research, he suggests that publicizing the fact that stereotyping is widespread—a message liable to be perceived as "most people do it"—might reinforce and perpetuate unwanted stereotyping.[111-113] Al-Gharbi emphasizes that people tend to remember *mis*information after repeated exposure to it, and that anti-bias training might inadvertently harden stereotypes.

While attempts to overcome biases may be well-intended, they may be clumsy. For example, attending a medical errors conference in Bern, Switzerland, in 2018, I recall an agreeable if eager-to-please British GP conveying the importance of bespoke bedside manner: "I have lots of Irish patients," he told the audience. "Since the Irish enjoy 'the craic,' I incorporate banter into their visits." ("Come back with a hop, a skip, and a jump for your MRI"?)

Education about unwanted biases is important. However, its real value may not be in directly improving patient outcomes but in increasing awareness of the many challenges involved in ensuring consistent, equitable care for all patients during face-to-face visits—regardless of race, class, creed, or symptoms. Unfortunately, exaggerating the effectiveness of short training programs could risk engendering a glib form of false consciousness. In short, there is no persuasive evidence that such strategies genuinely help to mitigate discrimination, communication breakdowns, or medical errors in patient care. As currently propagated, such programs might present the right optics at the expense of exploring more robust solutions.

The "art of medicine" resembles a complex, abstract painting—elegant in its intention, yet chaotic in its execution. In the next chapter we will consider whether manufactured "digital doctors" can introduce more precise brushstrokes.

8

Building Digital Doctors

Bill Gates, Mark Zuckerberg, and Steve Jobs abandoned college to revolutionize the world. In 2023, a new kind of prodigy shattered the mold of the dropout tech titan. Enter "GPT" (Generative Pre-trained Transformer model), an AI whiz that bypassed the traditional classroom altogether, yet effortlessly aced college exams.

Headlines buzzed about the first version of this digital marvel—"ChatGPT"—that seemed to reshape our perceptions of intelligence and learning. The bot beat 90 percent of lawyers, acing the Bar exam. It sailed through the written exams for the MBA program at the University of Pennsylvania's globally renowned Wharton School of Business.

In the world of healthcare, ChatGPT was holding its own in the gold-standard of American medical exams, without devoting a single second to traditional study. In February 2023, it scored between 52 and 75 percent in the three-part Medical Licensing Exam ("USMLE"), which has a threshold pass rate of around 60 percent each year.[1] This early ChatGPT wasn't top of the class, but the bot was getting better. In Japan, Tatsuya Haze and colleagues challenged the bot using the country's 2022 national medical examination. GPT-3.5 answered 56 percent of questions correctly but the later version—GPT-4—aced the test with a score of 81 percent.[2]

ChatGPT wasn't just triumphing in multiple-choice tests. Dr. Adam Rodman, an internist and researcher based at Beth Israel Deaconess Medical Center, tested ChatGPT using complex medical cases from *The New England Journal of Medicine*. The bot correctly identified the diagnosis 64 percent of the time and ranked it as the most likely diagnosis 39 percent of the time.[3] Not bad for a bot that never entered, never mind graduated, medical school.

Upping the ante, and using complex, real-life clinical cases—in other words, not the cosmetically tweaked or fabricated case studies found in exam papers or published in journals—a team from Denmark led by Alexander Eriksen undertook a study using the written words of real patients.[4] As Eriksen and his colleagues noted, this meant "long, complicated, and varied patient descriptions." They wanted to see how well ChatGPT-4 fared against humans in 38 clinical cases and compared the bot's answers to those offered by medical journal readers using nearly 250,000 online answers. The bot correctly diagnosed an average of 22 of the 38 cases (57 percent). Readers achieved an average of 14 correct cases—a modest 36 percent. In other words, ChatGPT-4 performed better than 99.98 percent of medical readers.

ChatGPT and its cohort of AI bot buddies look like languid geniuses. In this chapter we will go behind the headlines to ask how this algorithmic arriviste, these bots without brains, pulled off such impressive results. We'll explore whether today's digital tools possess the power to silence the overwhelming noise in medicine. We'll also ask if machines can surpass what doctors proudly call their "art of medicine."

Algorithmic expertise

The notion that a perfect clinician exists reveals more about our cultural myths about doctors than the reality. As we discovered, from tiredness to the time of day, from rookie training to the problem of aging, from mixed ability learners to gender and geographical location, there is a veritable clinic cacophony when it comes to differences in doctors' judgments. Variety in clinician aptitude isn't akin to the "spice of life," it's akin to a "spice of risk." A particularly voluminous source of noise is medical expertise. For very human reasons, doctors are confronted by a multitude of challenges in acquiring, sustaining, and applying medical knowledge. Moreover, these struggles span their entire careers.

Can bots quell the cacophony that interferes with care?

Let's start by examining how large language model (LLM) chatbots such as ChatGPT learn. There are two key methods. One route is via a kind of tech auto-didacticism or what is called "self-supervision." Here, fed high volumes of data, the machine works out patterns without any human interventions. Commercial LLMs such as ChatGPT are trained via self-supervision on publicly available datasets such as Wikipedia, Reddit, and other open-source online resources. Despite machines doing the lion's share of the work in discerning patterns in the data, strictly speaking even this kind of learning isn't wholly devoid of human input. This is because humans make prior decisions about what data troughs or "training sets" to feed the models—including whether data derives from social media platforms, articles published in online libraries, or wherever.

A second route relies more heavily on human tutoring, and this is where the moniker "machine learning" is truly a misnomer. Like Toto pulling back the curtain on the Wizard of Oz and exposing the real mechanics behind the supposed magic, there is a fundamental role for humans influencing bot "brainware." This human coaching is called "instruction tuning." It occurs in two main ways. One method is when chatbots-in-training are provided with instructions—for example a query—and humans supply the model with the kinds of responses they'd expect it to churn out.

A second, related method arises via so-called "reward-based tuning." Here the human instructor "rewards" the bot when it offers correct responses, "penalizing" it when it comes up with unexpected or wrong answers. Again, this nudges the weightings of its underlying architecture to inch it toward more desirable responses.

As a result of all these processes, LLM-powered chatbots can become amazingly adroit at recognizing, summarizing, and generating content. In summary: as for doctors, so for bots, the quality of the *training sets* matters—that is, the kinds of information bots are fed. Also, just as for human doctors, the quality of the *human coaching* matters too—something we'll return to later.

Consider, again, chatbots' medical exam results. It's important to emphasize that what makes their grades particularly impressive is that LLM chatbots such as ChatGPT, Bard, and Bing AI were simply not designed to undertake medical tasks. These bots didn't come close to entering the techy equivalent of machine-learning-medical-school—they are entirely untrained in medical matters, relying solely on the general LLM structure to produce their answers. Danielle Ofri lamented that medical students are stuffed with medical information like geese for foie gras, but these chatbots were fed a varied diet that included a junk food smorgasbord such as social media (e.g., X/Twitter, Reddit, Facebook), as well as online books, news media, and publicly available images.

Despite bots' bingeing on the learning equivalent of a greasy all-you-can-eat buffet, their medical test results were solid and improving. LLMs rely on vast numbers of units or parameters to embed knowledge. At the time of writing, GPT-3 reportedly has around 175 billion parameters, and GPT-4 is around six times larger: the system is said to be based on eight models each with 220 billion parameters, amounting to a total of around 1.76 trillion.[5] The increased volume of parameters has undoubtedly improved GPT-4's accuracy.

Aside from brute computational power, AI researchers emphasize that if bots are to do better, they need to be fed a more nourishing medical diet. This means quality-control training on clinically relevant materials that may not be publicly available. For example, it could include access to the mountains of articles held behind paywalls in journals such as *The New England Journal of Medicine*, as well as information within textbooks, and curated in clinical databases. Signaling that the quality of data matters, Japanese researcher Tatsuya Haze and his team found a positive correlation between the number of open-access medical papers fed into ChatGPT and the accuracy of its responses to the Japanese National Medical Examination; equally, the lower the volume of academic papers in any field, the more errors the chatbot made.[2]

Once again, it is crucial to consider that these bots were not designed to undertake the task of diagnostics. Yet they displayed expertise and clinical judgment via other means.

Robust, resilient bots

Bots defy traditional doctor stereotypes—they don't don classic white coats with stethoscopes draped round silicon necks—but this also means they are remarkably resilient. Machines muffle the noise associated with the vast spectrum of diversity among doctors. This doesn't imply that every source of noise is silenced. For example, chatbots are not always consistent. Consistency can partly be improved by the bot being fed the same verbatim question—or what AI researchers call "prompts." In real life, however, humans may want to ask bots a question in multiple different ways.

To solve this challenge, it could be that doctors and patients learn what works best when "talking" to chatbots to optimize their consistency. Studies show that so-called "prompt-engineering"—carefully crafting the right inputs—can improve how these models leverage their abilities.[6] Of course, during face-to-face medical appointments, patients often "attune" their own inputs too—Erving Goffman called it "the presentation of self in everyday life."[7] Social media is replete with advice, including sartorial action ("dress up to be taken seriously") or pre-visit planning ("bring notes into the appointment so you don't forget, even when they interrupt you"). It could be that using chatbots sometimes demands a different kind of preparation; whether this proves more burdensome for patients, or doctors, remains to be seen.

Doctors differ widely in their clinical competency. Commercial chatbots are not uniform either, but the spread of variation is considerably less. One study in pharmacy found that on average ChatGPT was more reliable in its recommendations than 20 expert clinicians by offering consistently accurate responses.[8] In contrast experts' responses hinted at the unwelcome rainbow of diversity in healthcare. While averaging an accuracy of 69 percent, clinicians' accuracy scores ranged from 48 percent to 94 percent.

Machines silence other sources of nasty noise. Being devoid of brows and brains, bots don't sweat or stress. They are incapable of being fatigued or hampered by fleshy constraints. Nor are they hostage to circadian rhythms, low blood sugar, or distractibility. Again, this isn't to suggest bots are always right, but their brute computational power certainly does have potential to hush the hubbub of healthcare.

Consider, for example, the noise associated with doctors' learning. Physicians have barely any time to read, never mind look at the latest research findings, and need to sleep daily. In contrast, machines crunch their way through open-source data at breakneck speed, without needing to stop for a breath or a break, or even to pee.

Like speed-freak bookworms, AI has a stunning capacity to ingest medical publications and data in seconds, and round the clock. They make doctors munching through a medical journal look like a quaint case of the children's book *The Very Hungry Caterpillar*.

Doctors and commentators rightly pointed out that the inaugural version of ChatGPT, rolled out in November 2022, was only trained on web text published up until September 2021. While this is true, it overlooked the tough realities of how humans acquire and transmit medical expertise. ChatGPT's initial two-year backlog was not close to the snail-paced 17-year timescale of "bench to bedside" that it still takes for medical research to dribble into doctors' practices.[9] Moreover, denting the idea that ChatGPT was worryingly out of whack, by mid-2023 the complaint was already old news.

Alongside the noise associated with keeping abreast of medical advances are the challenges of committing to memory vast amounts of information. As a result of this truly impossible task, doctors' practical expertise is narrower and more parochial than patients need it to be.

To pluck one example, many women with endometriosis wait years before receiving a diagnosis. Because it can rapidly hoover up information on the condition, one study found that ChatGPT accurately responded to more than 90 percent of questions about endometriosis, inviting the question: How well would the average primary care doctor do?[10] Someone who suffers with the condition informed me that, on the cusp of menopause—just as nature was about to solve the problem—a doctor finally diagnosed her. Of course, this is only "anecdata"—a personal story and not a hard factual generalization. Yet doctors suffer from anecdata too. Struggling to update and recalibrate their expertise, they draw on the finite number of patients they have seen during their careers ("I've seen this disease in 10 patients before").

In contrast, AI can base its insights on millions of patients from all over the world. As Stanford University medical legal scholars Michelle Mello and Neel Guha observe, "The risk of getting a wrong answer may be lower from model output based on hundreds of thousands of sources—even if some of the sources are unreliable—than from asking one colleague."[11]

The people who may have the most to gain from this raw processing power are those with medically rare diseases. AI's special talent is in seeking out hidden patterns in data. As Harvard Medical School researcher Andrew Beam says, bots like ChatGPT work like a "super high-powered medical search engine."[12] Because it has devoured the entire internet, ChatGPT can do a pretty decent job of detecting atypical patterns of symptoms associated with rare diseases.

DIAGNOSIS AND TREATMENT

It might even do it better than doctors. Exploring its rare disease smarts, Austrian researchers entered 50 clinical cases including 10 rare illnesses into ChatGPT-4. The bot came up with the correct diagnosis for common conditions within two suggestions, solving 90 percent of the rare illness case studies within eight suggestions, easily outsmarting the doctors enrolled in the study.[13]

One such patient who benefited from AI's talents is Alex, a little boy who saw 17 doctors over three years for his chronic pain.[12] From orthodontists to pediatricians and from neurologists to internists and even a musculoskeletal specialist, Alex's mother Courtney tried them all. No diagnosis or treatment plan seemed to fit. Exhausted and exasperated, Courtney plugged Alex's medical notes into ChatGPT which eventually suggested the possibility of "tethered cord syndrome" (TCS), a rare condition closely associated with spina bifida, where part of the spinal cord doesn't fully develop.

In the US, TCS affects fewer than 1 percent of the population. Like Alex, children with this condition experience numbness, pain, muscle weakness and problems with motor control. Securing diagnostic confirmation from medical specialists, Alex is now receiving suitable treatment. Courtney told journalists that thanks to ChatGPT she now feels "every emotion in the book, relief, validated, excitement for his future."

We discovered that when it comes to the "art of medicine," doctors learn to detect patterns of symptoms, and convert this to automatic, intuitive "System 1" knowledge. Undoubtedly this is an impressive human feat. However, it pales in comparison to the mind-boggling potential of AI to discern complex patterns and relationships in data. AI holds promise in detecting markers of rare diseases—via biological test results, or genetic signatures—even before patients experience symptoms.

A team at Harvard Medical School and Massachusetts Institute of Technology trained a model of AI which they called "SHEPHERD" to explore these hidden connections.[14] The AI was designed using simulated patients—that is, data devised to be similar to that of real patients but which protects their medical history by avoiding any real-life information. Investigators tested the model using nearly 500 real patients with 299 different diseases, including exceptionally rare genetic illnesses. After herding a massive volume of information, SHEPHERD proved exceptionally agile at rounding up the genes causing rare diseases. On average, it found the right gene after three or four attempts. The authors concluded that SHEPHERD could help to "shorten the diagnostic odyssey for rare disease patients."

For patients like Dr. Keith Geraghty—the Irish academic we met who lives with ME—AI is already beginning to shed light on the biological markers of this illness. While Keith's symptoms were treated as a mental health issue by his doctors, AI has identified immune cell abnormalities that many patients with ME share.[15] In the future, AI diagnosticians might help patients avoid the medical trauma associated with the kinds of diagnostic errors people with ME have faced.

AI can improve efficiency in clinical trials. A model called TrialGPT slashed patient screening and recruitment time by 42 percent.[16] AI can also act as a stand-in for control groups in clinical trials by offering information about how diseases usually progress, and predicting how drugs might interact with different biological pathways. By rapidly analyzing vast amounts of genetic, biological, chemical, and clinical data, it can identify new candidate treatments and offer suggestions for repurposing drugs that are already on the market.[17]

This isn't an academic pipe dream. Matt Might is a professor in computer science at the University of Birmingham, Alabama. For Prof. Might, rare disease is personal. He used his expertise in AI "to hunt"—in his own words—"his son's killer."[18]

A pioneer in the field of "precision medicine," Matt Might used genomic sequencing and applied his skills in computational biology and data analysis to dive deeper into his family's genetic data. Using data-mining techniques, he determined that his son's killer was a mutation in the "NGLY1" gene. Matt's efforts led to his son Bertrand being identified as the world's first known case of NGLY1 deficiency. Bertrand lived till he was 12 years old, and during his life his father used AI to sleuth out and to tailor a variety of treatments that helped alleviate the severe symptoms associated with Bertrand's condition. Although Bertrand's life was short, it was not in vain. Matt told me, "His life was pretty powerful."

Driven by his care for Bertrand, his father established a global patient community for rare diseases. As for NGLY1 deficiency, Matt says: "Today, we're on the order of about a hundred identified cases across the world. So, it's still super rare, but there's at least a sizable community now. There's even a gene therapy and clinical trials which started in January [2024]."

He adds: "We've come a long way for a disease that a decade ago doctors didn't know existed."

Algorithms, bots, and biases

Serena Williams, Keris Myrick, and Keith Geraghty believed they were not being taken seriously by their doctors. If patients can access the technology—a big *if* explored earlier—chatbots will "talk" to anyone regardless of sex, income, education, skin color, or ethnicity. Bots don't judge us on how we present, nor do they discriminate against body size. They are incapable of interrupting, withholding information, showing disdain, or displaying social distancing. Chatbots are blind to the things that humans cannot unsee or unhear. Yet these benefits are rarely mentioned in medicine.

Emphatically, however, this doesn't mean bots are bias-free. While doctors' education is riddled with errors and omissions causing unfair treatment, how AI is trained can also embed unwanted biases—giving rise to the slogan "garbage in, garbage out." When unwanted biases are "baked" into machines leading to unfair recommendations, this is called "algorithmic discrimination."

Starting around 2016 with Cathy O'Neil's book *Weapons of Math Destruction*,[19] greater attention has been paid to this form of discrimination. Troubling racist and sexist responses have been found in all kinds of high-stakes decisions where algorithms have been used, including in facial recognition software, criminal sentencing, employment, and housing allocation. A significant portion of this research has been, and continues to be, led by Black and female scholars, a fact that is not coincidental.

In medicine, there is huge scope for machines to perpetuate unfair treatment via coded biases. Many health systems rely on commercial algorithms to decide which patients with complex health problems might need more help. The accuracy or otherwise of these approaches has the power to affect millions of people. For example, in March 2024, an independent review conducted by the UK government found gender and racial bias in medical tools and devices.[20] We've seen that the diversity of patient populations matters for the validity and effectiveness of these tools; however, a review found that more than half of the data used in health AI originates in the US and China.[21]

The linguistic competence of chatbots also needs to be more thoroughly investigated. More than half of all websites are published in English, yet worldwide, fewer than one in five people speak it. Arabic is the fifth most widely spoken language, yet in 2023, an early study found that ChatGPT's responses to medical questions were 33 percent less accurate in Arabic than English.[22]

Things are certainly improving. In 2025 one study found that ChatGPT-4 accurately analyzed medical notes in Spanish and Italian.[23] But languages with the fewest speakers, which are often the most marginalized online, continue to have disproportionately limited digital representation. Even among English speakers, those who don't speak standard American English may find bots culturally clumsy on occasion, and questions remain about AI competency when it comes to accents, local vernacular, and colloquialisms.

How AI tools are designed also has repercussions for how well they work. Dr. Ziad Obermeyer, a trained medical doctor and health informaticist at the University of California, Berkeley, led a study investigating how risk scores for primary care patients were managed at a large American academic hospital.[24] His team found that when the algorithm tried to figure out who was most sick, it looked at how much money was spent on their healthcare. Black patients often have less spent on their healthcare services despite the fact their needs are greater. Dr. Obermeyer's team recognized this proxy was likely to be unfair because it assumed Black patients were less sick than they truly were. Sure enough, they found that making the algorithm fairer could increase the percentage of Black patients getting help from 18 percent to 47 percent.

Probing the potential for racism in clinical responses, a study by a team at Stanford asked nine clinical questions of the commercial bots ChatGPT-3.5 and GPT-4, Bard (now Gemini), and Claude. While every bot they studied acknowledged that race was a social construct, the devices sometimes contradicted themselves by offering essentialist biological explanations of race.[25] For example, asked about kidney function, an area of medicine with race-based practices, ChatGPT-3.5 and GPT-4 justified their responses by saying Black people tend to have higher creatinine levels because they have different muscle mass than whites; while the former tends to be true in the US, the latter is recognized to be factually incorrect.

In another study of ChatGPT-4, looking at conditions with similar rates of prevalence across all patients regardless of gender—such as COVID-19 and colon cancer—the model was "substantially" more likely to generate cases of men having these illnesses than women.[26] When it came to treatment recommendations, the bot was significantly less likely to recommend advanced imaging—CT, MRI, or abdominal ultrasound—for Black patients compared with whites. The bot also rated white men as significantly more likely to exaggerate their pain level compared with Asian, Black, or Hispanic men and women.

Are bots sometimes prejudiced? The short answer is "yes." AI reflects individual and societal prejudices, including those perpetuated within the medical profession. Algorithms aren't autodidacts—they learn from their human masters. AI is also heavily reliant on the quality of the historical data fed into the algorithm. If medical chatbots are trained using doctors' inappropriately biased decisions, such as those documented in clinicians' notes housed in our electronic health records, this can also bake in bias. And if, like doctors, they devour the same menu of skewed publications, textbooks, health records, and other clinical data, this perpetuates medicine's failings via the same original sources. Other human choices can infiltrate technology too. During training phases, if the AI receives incorrect feedback from doctors or other experts in the loop, then the same bad habits and shortcuts can become ingrained in algorithms. Consequently, the miseducation of machines means chatbots can very easily become prejudiced.

Debiasing AI for better care

Most attention is on what bots do badly. Attesting to this, in 2024 a study led by Emma Pierson, a computer scientist at Cornell University, found that 85 percent of papers which examined the effects of LLM chatbots on equity focused on the potential for harms, but only 15 percent considered the opportunities.[27] As a result of this focus—or let's call it what it is: bias—some have even made the inference that, because of algorithmic discrimination, we are better off relying on humans, rather than dabbling with technology.[28]

By fixating on AI's flaws, researchers have exposed an almost romantic inattentiveness to the scope for discriminatory biases perpetuated by doctors themselves—not just in terms of what physicians learn in medical school, but what they do during face-to-face appointments with patients. Furthermore, it is futile to scold an inanimate entity for its shortcomings. Doing so only shifts responsibility away from humans. Ironically, tendencies to blame bots via some sort of outgroup, anti-AI bias might even be said to illustrate, in a novel way, the very prejudice that we profess to abhor.

Quite simply, the rush to favor human doctors over AI is moving far too quickly. It turns on two hasty assumptions. The first, as we've noted, is the ultimately false assumption that doctors are somehow less guilty of discrimination, or are better able to manage it when it does arise. The second is that AI cannot be used to strengthen equity in healthcare. Once again, if a key function of medicine is fair treatment, it is critically important that we ask who *or what* can do a better job of delivering on this.

Fortunately, researchers are increasingly recognizing that technology can offer reasons for optimism. Instead of *idées fixes* about the limitations of AI, these researchers recognize that AI helps hold a magnifying glass to human biases, and can be deployed to improve fairness.[29,30] Although there is as yet no standard guidance on how to reduce unwanted algorithmic biases, in this respect it is no worse off than human-mediated care, which also lacks effective, evidence-based means to overcome unwanted stereotyping.

Addressing historic omissions and biases in publications fed to AI models is crucial. Improving recruitment and ensuring diverse patient demographics in training data can enhance representativeness. Testing algorithms on varied populations, refining prompts to boost performance, and openly considering what fair outcomes look like are also key steps. With these processes in place, AI has the potential to reduce biases in medicine more effectively and swiftly than traditional methods.

Already, research hints at the potential. Because it represents data as probabilities rather than discrete categories, AI also has the ability to offer more nuanced interpretations of how racial and other identities might be associated with health outcomes.[31] For example, a study by researchers at Harvard Medical School examined how well the AI model GPT-4V, which can understand both text and images, performs on difficult medical cases. The team tested it using 934 medical cases from *The New England Journal of Medicine* Image Challenge, spanning from 2005 to 2023, and compared its performance to doctors' responses. Their results showed that, with an accuracy rate of 61 percent compared to 49 percent for doctors, GPT-4V easily surpassed clinicians. Regarding the 420 cases that included pictures showing different skin tones, GPT-4V was better than doctors at accurately diagnosing conditions across all skin tone categories.[32]

Another source of discrimination in healthcare is doctors' thought processes. We learned that doctors' expertise relies on nonconscious, automatic "System 1" thinking. This means doctors are unaware of the underlying processes guiding their decisions, making it difficult for them to recognize prejudice and unfair stereotyping when it arises in their practice. Despite medical *mea culpas* and must-do-betters that may ease doctors' consciences over discrimination, as we discovered, there is scant evidence that anti-bias training helps the patients who bear the brunt of unfairness.

Fortunately, people are better at debiasing others than they are at debiasing themselves, and this means they may do a better job of challenging algorithms too. AI can even be used to monitor when institutions are missing data for patient populations, when humans misrepresent research findings, or even to discern patterns of discriminatory language used by doctors in our records. Those studies that we highlighted in the last chapter that uncovered stigmatizing language in the medical records of African Americans, female patients, and those with "less prestigious" conditions relied on AI's unparalleled ability to analyze thousands of notes at speed, effectively holding a mirror up to the behavior of medical professionals.[33,34]

Although the benefits could be considerable, the path will not be straightforward. Throughout the book we've seen that there are real, biologically meaningful reasons why people from some backgrounds are more likely to fall ill, including higher risks of social and environmental stressors, or heightened vulnerabilities to genetic illnesses.

DIAGNOSIS AND TREATMENT

The causal links underlying many of these risks are not always well understood. Undoubtedly, some proxies—such as race or ethnicity—are far from perfect when it comes to tracking disease risks, but because many societies haven't ignored race or ethnicity, there are times when it may be harmful for medicine to overlook these social categories too.[35] It's early days but there is evidence that AI could be used to tackle biases in ways human doctors simply cannot.

As a teenager, Emma Pierson learned that her mother carried the BRCA1 mutation which confers a high risk of breast and ovarian cancer. Since her mother was Ashkenazi Jewish, she was tested, though Dr. Pierson recognizes ethnicity functioned as an imperfect proxy for the information that doctors would ideally need—namely, her mother's detailed genetic ancestry.[36] Despite these limitations, Dr. Pierson says: "I still believe the doctors made the right call. My mother was 10 times as likely as women who aren't Ashkenazi Jewish to carry these mutations, which increased her risk of ovarian cancer by a factor of 30." With a ten-year survival rate of 36 percent, Dr. Pierson emphasizes, "Had the doctors not tested her, she might not have lived to see us grow up."

In her day job as a data scientist, Dr. Pierson acknowledges the need for much more nuanced data and techniques to directly pick up on the underlying reasons for demographic differences in disease prevalence. But she is pragmatic. Proxies can be better than nothing.

Research reveals the razor-edged challenges of this tightrope walk. On the one hand, the inclusion of patient race can degrade clinical accuracy and healthcare decisions.[37] On the other hand, algorithms that omit race can under-predict the cancer risks for African American patients.[38,39]

Careful evaluation, monitoring, and continuous auditing of algorithms will be required to ensure the right decisions are made. For this we need researchers to become more explicit about what is meant by "fairness"—how algorithms impact people from different demographic groups, and whether it is always helpful to think of patients in terms of "group" membership as opposed to their uniqueness as individuals. These decisions carry huge consequences for diagnostic accuracy, as well as financial and occupational stakes, including patient eligibility for medical care, and disability compensation.[40]

Machines also have unique traits of their own. AI can go well beyond what we humans *think* we might be teaching it to do. Dr. Judy Wawira Gichoya, a health data scientist and radiologist at the Emory University School of Medicine in Georgia, found that AI can accurately predict patients' race from medical images such as X-rays, CT scans, and mammography. This AI patient-identity superpower was achieved without any explicit demographic labeling of the data. Even when a range of proxies for race were removed—the sorts of clues that the AI might be picking up and using as shortcuts—it consummately managed to work out patients' racial identities—something that specialist doctors can't do.[41,42] The exact way AI is doing this isn't clear, but these models could be identifying signs of other diseases that people from some demographics are more vulnerable to acquiring, and that are subtly manifested in the images.

Is this a superpower that we want or need? If a patient's race is identified without their confidential disclosure, especially in the wrong hands it could result in harm or exploitation, and potentially exacerbate prejudice.[31] Dr. Gichoya recognizes this; however, she argues that, if sensitively handled and with appropriate ethical oversight and guardrails, this puzzling AI superpower could be harnessed for good. This is because it could help clinical researchers to better track disparities in care, to highlight a wider variety of hidden inequities, and to identify the health burdens faced by some populations.

Take pain, a phenomenon that is unequally distributed in society. Doctors sometimes discriminate against patients, discrediting their pain during face-to-face appointments. AI could be superior to humans at gauging pain experiences among people from different backgrounds. Dr. Emma Pierson and her colleagues used AI to assess the severity of osteoarthritis by looking at patients' knee X-rays and predicting how much pain they might feel.[43] Radiologists could predict about 9 percent of pain differences among patients from different races, but AI predicted 43 percent—almost five times more—with similar results for patients with low incomes and those with fewer years of education. For patients who are suffering, AI could be revolutionary.

Is it time for doctors to divest themselves of their white coats, hang up their stethoscopes for good, and pass the prescription pad to bots? Before we examine this question—which is not as easily answered as some might suggest—let us dig into some subtle yet striking problems with AI.

Bots bumbling in the dark?

Dr. Jeremy Faust is an emergency doctor at Brigham and Women's Hospital in Boston and Editor-in-Chief of *MedPage Today*—one of America's biggest medical news sites. Back in January 2023, Dr. Faust was among the first doctors to explore the use of ChatGPT.[44] In his own words, he was "blown away" by the "astounding" capabilities of the chatbot. But, unlike his ambitious namesake—the Dr. Faust of German legend who eagerly traded his morals for unlimited powers—this Dr. Jeremy Faust resisted making a pact with the "devil" or in this case a commercial AI company.

Faust uncovered some suspicious findings. The computer told him that costochondritis, a common cause of chest pain, can be caused by oral contraceptive pills. Having never heard of this risk, he asked the bot to back it up with some evidence: "OpenAI came up with this study in the *European Journal of Internal Medicine*. ... I went on Google and I couldn't find it. I went on PubMed and I couldn't find it. I asked OpenAI to give me a reference . . . and it spits out what looks like a reference. I look up that, and it's made-up. That's not a real paper." Soon other doctors and journalists raised red flags.[45]

LLM bots can be fibbers, frauds, and far from the brightest bulb in the room. Like a computer-processing Pinocchio, they make ballsy blunders that sound compellingly correct. This tendency is referred to as "hallucinations." In humans, hallucinating is what happens when we perceive things that are not actually there. In chatbots it means AI's tendency to generate statements that are probabilistically plausible given the data it has been fed, but are false.

Research piled in. One study found the rate of bot bloopers was pitiful: requesting journal articles on a variety of medical topics, investigators found GPT-3.5 spat out fake publications 98 percent of the time.[46] While newer model GPT-4 produced a significantly lower percentage of made-up medical articles—at 21 percent—investigators modestly noted that the error rate "remains nonnegligible." Some confabulations were so sophisticated this led to rigorous checking in online libraries, with investigators even writing to the alleged authors—some of whom actually did exist—to verify whether the erudite-sounding scientific papers were indeed published (they weren't).[47]

ChatGPT's summaries of medical information could also be subpar. In a study led by Liyan Tang at the University of Texas, Austin, researchers asked ChatGPT to abridge findings from the highly respected and rigorous Cochrane reviews of medical evidence. Aside from making stuff up, the bot sometimes misinterpreted information, contradicted "itself," or was unjustifiably certain about clinical evidence.[48] It also tended to make more mistakes with longer pieces of text.

From a patient's perspective, this is profoundly troubling. To err is human, so why do bots make blunders? These tools are essentially "stochastic parrots"—the phrase was coined in 2021 by Dr. Emily Bender, a professor of linguistics, Dr. Timnit Gebru, a former co-lead of the Ethical AI team at Google, and their collaborators.[49] "Stochastic" refers to phenomena that are randomly determined. Just as a parrot can mimic human speech without comprehending its meaning, AI models parrot back information without a true understanding of its content. Due to their design, this AI can't distinguish true from false, gets confused by rephrased questions, struggles with negations like "no," "not," and "never," doesn't understand causality or simple math, and lacks basic common sense.

While the frequency of these fabrications is declining with advances in AI models, errors still arise. Examining information on cancer treatment offered by ChatGPT-3.5, Harvard researchers found that 13 percent of outputs were hallucinations.[50] More recently, another study in radiology found that ChatGPT-4 demonstrated strong performance but occasionally offered recommendations—in 2.8 percent of cases—without drawing on direct evidence.[51]

Or consider "ambient AI," which quietly "listens" during our visits and transcribes doctor–patient conversations in real time, with the goal of reducing physician paperwork. Extremely worrying, however, in October 2024, a University of Michigan study found that OpenAI's Whisper, used in the Nabla ambient AI transcription tool by 45,000 clinicians in over 85 US health organizations, produced fabricated text in 80 percent of transcriptions.[52]

These limitations, and the fact that bots can be extremely bright in some ways but asinine in others, are regularly discussed by cognitive scientist and AI commentator Gary Marcus. A techno-enthusiast, yet currently an ardent AI skeptic too, Marcus argues that current AI exhibits "narrow intelligence"—it excels at just one thing: pattern recognition. AI models, he says, are "digital idiot savants," with chatbots such as ChatGPT functioning as a "kind of glorified cut and paste" engine.[53,54] Such pattern-crunching, says Marcus, does not substitute for real-world understanding. Moreover, he argues that marinating these models in even greater troughs of data or devising stricter training regimes will not be sufficient to overcome their current limitations. In his view, this will only provide more statistical probabilities instead of truly knowledgeable insights.

If these models lack genuine self-insight, they also lack penetrability. This latter phenomenon is known as AI's "black-box" problem, referring to the lack of transparency and interpretability in how algorithms generate their responses. Beyond proprietary concerns about AI, accessible only to insiders, their internal workings involve complex systems with millions of parameters. Humans lack the brute brain power to grasp how inputs are transformed by AI into outputs. Moreover, AI's complexity and scale give rise to "emergent" talents—unexpected abilities like solving problems it wasn't explicitly trained for. This makes AI something of an alien intelligence that our minds don't fully understand.

With all their serious faults, bots don't drop into some pristine medical bubble, however. Bots and docs make blunders in different ways. But both do it. Doctors seek the truth, but rather than hallucinate they confabulate. Proffering explanations for their own judgments, doctors provide idealized pristine, post hoc analyses—via System 2's salesmanship—rather than perfect traces of the causes behind their expert System 1 intuitions. System 2 is an apologist for doctors' System 1 thinking rather than a perceptive critic of it, as we discovered.[55]

The question is, who or what is worse? Many AI studies overlook the comparative performance of human doctors. Bucking the trend, in the Cochrane GPT study, Liyan Tang observed that human-generated summaries contained "a higher proportion (28 percent) of fabricated errors, resulting in more factual inconsistency and potential for harmfulness in references."[48] Doctors routinely make serious errors too, including in documentation. As discussed earlier in the book with Dr. Allen Wenner, patient documentation is often completed long after visits, increasing the risk of mistakes. Doctors also tend to "copy, cut, and paste" from one part of patients' electronic records into other areas, without careful checking. So, while AI introduces errors, doctors do too. Head-to-head trials between AI and doctors are revealing, and we need more of them. In one study, ChatGPT-4-generated discharge letters were comparable in quality to those written by junior clinicians and showed no hallucinations.[56]

The emergent field of "explainable AI" strives to render machine processes more visible.[57] Fascinatingly, even when AI functions as a stochastic parrot, it seems to do a pretty good job at offering plausible explanations for its clinical reasoning. In 2024, Dr. Adam Rodman again led another cutting-edge study investigating exactly this capacity.[58] His team put GPT-4 to the test, exploring its ability to reason through patient cases step by step. Using 20 clinical cases, they compared the bot's performance to that of internal medicine residents and seasoned attending physicians in Boston.

The AI not only outperformed both groups in summarizing problems and clinical reasoning but did so with greater consistency. It even matched doctors in diagnostic accuracy. ChatGPT-4's reasoning errors were more frequent than those of residents, but its overall error rate was comparable to that of the experienced attending physicians. In short, the tool wasn't perfect, but it was strikingly competent—offering a glimpse of what AI might bring to medicine's diagnostic future.

OpenAI's o1 and o3 and DeepSeek's R1 series may represent a new kind of model that challenges the charge of "narrow intelligence"; though many are skeptical of the hype. This latest AI uses "chain of thought" reasoning, a method where the model breaks down complex problems into clear steps, searching through different responses to arrive at an answer. While too early to call, this approach could take AI to new levels of clinical accuracy, and we'll explore early findings in the next chapter.

It might be objected that, regardless of the incarnation, AI is merely exhibiting ersatz reasoning rather than the real thing. Yet AI's opacity and its own imperfect reconstructions of its reasoning should not be held to a higher bar than that of humans. While current bots bungle blindly in black-box bafflement, oblivious to their own operations and origins, medicine harbors its own dark arts that generate errors. Even if machines lack transparency and interpretability, clinicians' expertise can also be a chaotic enterprise.

Moreover, AI can even illuminate doctors' decisions. Dr. Ziad Obermeyer used machine learning to analyze how doctors diagnosed heart attacks by comparing their decisions to those of a predictive algorithm.[59] His team discovered that doctors often over-tested patients who were at low risk of heart attacks, while failing to test high-risk patients—some of whom died, after being overlooked by physicians. Publicizing his findings on X/Twitter, Obermeyer commented, "BTW: we're not studying some random docs—They are 100% residency-trained emergency physicians, at a top-ranked academic medical center. Full disclosure: I was one of them! It was eye-opening and humbling to see my mistakes. What was I doing wrong?"[60] It turns out human doctors oversimplified the nature of risk assessments, placing too much emphasis on prominent symptoms like chest pain. AI was holding the art of medicine to account and shining a light on physician error.

Although Bill Gates, Mark Zuckerberg, and Steve Jobs failed to complete their university degrees, we may not feel ready for doctors to do the same. AI may impress on exam questions and tests but there is much we still don't know about what technology can safely do.

That said, and although red-flagging AI risks is critical, it does not deal a decisive blow against the rise of and need for technology that "thinks." For thousands of years doctors have had the first and last word about medicine, and a growing body of research signals that human doctors no longer have a monopoly on medical expertise. While it is essential to avoid naïve trust in technology, we must also beware of the false facade of familiarity. Doctors frequently fail too when it comes to diagnostic accuracy.

The human "art of medicine" is marred by messy brushstrokes that do not do justice to every patient. Despite AI's limitations, it would also be churlish not to acknowledge that it is exceeding expectations. In the meantime, many assume that when humans and machines blend their talent and palettes, they can create a superior masterpiece in medical care. In the next chapter, we will challenge that assumption.

9

The Shotgun Marriage of Man and Machine

"My watch saved my life." These are the words of a 75-year-old retired teacher from Boston, Massachusetts.

Liam—who wanted his name changed for reasons that will become evident—is chatting to me from his ground-floor home office. The walls are fully lined with family photographs, guarded by multicolored liqueurs and artifacts gathered from his travels.

"Okay," Liam says with a jolt, "two years ago, my son-in-law gave me an Apple Watch which I loved. I could talk to my watch. I could find out what was going on in my body. I could even measure my blood pressure. And one of the things that interested me is that it sampled my heartbeat per minute—four times a night when I was sleeping, and at various points during the day. I could always check my watch to see what was going on."

Quite soon after wearing it, Liam noticed cardiac changes: "My heart rate was usually in the 60s, which is healthy. I swim regularly. Every so often my heart would beat at one rate and then at another rate, and then another rate, but then it would stabilize and go back. So, I didn't pay too much attention."

"But after a while it did become more and more obvious. And I could even hear my heart beating. It was loud, and I'd never experienced that before. My watch told me I had an irregular heartbeat. Sometimes my heart rate would be very, very irregular, beating well over the hundreds—actually 150, 180, nearly 200 beats a minute, which is not good. My heart was jumping around from 87 to 120 to 43. At one point it was 38—you're almost dead."

"So," says Liam, "I began to worry."

"Then my watch told me that I had atrial fibrillation."

Atrial fibrillation—often called "AFib"—is a common heart condition, and not normally serious, but in extreme cases it can cause heart failure. AFib occurs when the heart's chambers contract randomly, and sometimes so rapidly that the heart muscle cannot relax between contractions. The causes are not fully understood, but older people are more vulnerable, and the condition increases the risk of blood clots and stroke.

Liam immediately made an appointment with a cardiologist and explained what the watch was telling him.

"She said, 'I don't think you've got atrial fibrillation.' She had not touched me. She had not put her hand to listen. Instead," he says, "she focused on the findings of an EKG performed only a few months previous." More disturbing to Liam was what happened next: "As she walked out of the office she said, '*Oh, well, maybe you do have atrial fibrillation.*'" With a pained look, Liam adds, "I didn't know what to do."

Returning home, Liam's heart rate didn't improve. "I was just feeling worse and worse." And still his watch told him he had AFib.

Exactly one week later he "felt terrible." Rachel, Liam's wife, was adamant that he phone his nurse practitioner to explain the situation. "She told me, 'Get to the emergency [room] as fast as possible.'"

Liam was admitted without delay. "They examined me and said, 'Yeah, you've got atrial fibrillation, all right. A bad case of it.' So, they checked me into the hospital that night." Liam was treated with cardioversion, a therapy which involves shocking the heart and restarting it to restore what is called a normal sinus rhythm. "They wheeled me in, put me under, and did the cardioversion right on the spot. I woke up and my heart was in a normal sinus rhythm."

He adds: "Here's the thing, I can't forget: my watch diagnosed it earlier—much earlier—than the doctors."

It is often imagined that healthcare needs a new kind of conjugal relationship: a harmonious marriage between doctors and machines. The idea is that digital tools will not replace—or "disintermediate"— doctors. Instead, it is envisaged, these tools will help to augment and improve doctors' clinical thinking.

Equally, in wedding planning, it is often lamented that couples too often focus on the big day rather than their long-term compatibility. As English poet Alexander Pope once put it, "They dream in courtship but in wedlock wake." The same can be said about human–computer relations. This chapter interrogates the intricate history of doctors and digital devices. If man and machine is an arranged union, we will confront the sad reality that, already, many doctors may feel like filing for divorce.

We'll ask why Liam trusted his Apple Watch, why his cardiologist didn't, and how the dynamic between humans and machines might work better. We'll also get to the heart of the matter and ask if the current arrangement should be annulled altogether, and whether a new kind of clinical companionship is now required.

Doctors, nurses, and disrespect

The dynamics between doctors and technology is "complicated"— that old Facebook phrase for relationships infused with mixed feelings and unresolved issues. Before we become a fly-on-the-wall of clinics, discreetly observing the intimate interactions between doctors and AI, it is helpful to put clinical relationships into context. Earlier in the book we explored the sometimes-uneasy relations between patients and doctors, but we haven't yet probed what goes on behind the scenes between doctors, nurses, and trainees. Despite gleaming images of chlorine cleanliness, medicine's work culture isn't always as sanitized or as spotless as one might expect.

Consider a fascinating study conducted in 1966. Psychologists staged a series of telephone calls to 22 wards in different American hospitals asking to speak with the nurse in charge. The experimenter, who introduced himself as "Dr. Hanford," requested that the nurse administer 20mg of Astroten to a named patient on the ward.[1] Unknown to the nurses, Astroten was a made-up medication planted by the experimenters among the ward stock and clearly labeled:

ASTROTEN
5mg capsules
Usual doses: 5mg
Maximum daily dose: 10mg

This deceptive experiment opened four serious grounds for refusing to proceed: each nurse was requested to administer *a lethal dosage* of an *unknown medication* by an *unknown doctor* via *an unauthorized procedure*. Even the red flags were waving red flags. Yet, in secretly recorded phone calls, 21 out of 22 nurses had no hesitation in fulfilling the "doctor's" request. Investigators were struck by the "unmistakable" note of "courtesy and respect" of the nurses toward the so-called physician, behavior apparently triggered by the simple title "*Doctor.*"

Although the Astroten social experiment would not receive ethical approval today, it points to hierarchy and deference on hospital wards: the nurses failed to challenge the doctor. Research shows this is a common problem, and not only among nurses. A recent review of studies found "a palpable influence of the medical hierarchy" when it comes to reporting or speaking up about medical errors.[2] It found an estimated 60 to 70 percent of medical errors are communication related, with 23 percent due to failures of professionals to speak up.[2]

As with patients, deference and fear seems to interfere with candor. For example, one study found that the willingness to ask a doctor to complete hand hygiene decreased as the seniority of the doctor increased: 86 percent were prepared to ask an intern, while only 40 percent were prepared to ask a consultant.[3]

Writing about the muzzling effects of hospital hierarchy, British neurosurgeon Henry Marsh says, "one never openly criticizes or overrules a colleague of equal seniority."[4] Reflecting on "Morbidity and Mortality" meetings, where doctors are supposed to openly discuss mistakes and lessons are supposed to be learned, he observed, "the ones I have attended, both in America and in my own department are usually rather tame affairs, with the doctors present reluctant to criticize each other in public."

Medical pecking orders are also associated with bullying. In his bestselling roman-à-clef *The House of God*, Samuel Shem portrayed for a mass audience the grueling rite of passage that is medical internship. Set in a fictionalized version of Beth Israel Hospital, Boston, the semi-autobiographical novel published in 1978 is now a cult classic among medical students, though it was severely slated by senior physicians on its publication. Shem produced the first no-holds-barred account of professional hierarchies, painting a vivid picture of abuse and sycophancy. "The House medical hierarchy was a pyramid," he wrote, "a lot at the bottom and one at the top," but "Given the mentality to climb it, it was more like an ice cream cone—you had to lick your way up. From constant application of tongue to next uppermost ass, the few toward the top were all tongue."[5] Domineering behavior was rampant: "the intern was at the bottom of the other hierarchies," according to Shem: "In many tricky ways he had the opportunity to be abused at any time by Private Doctors, House Administration, Nursing, Patients, Social Services, Telephone, Beeper Operators, Housekeeping."[5]

The origins of this behavior will be familiar from our exploration of doctor deference, where we examined the relationship between doctors' prestige and dominance in patient interactions. Similarly, obtaining unrivalled rank ("prestige") through their skills, more senior doctors acquire immense capacity to bully ("dominate") those who depend on them—whether these junior staff are interns, residents, or nurses. This is no great secret among health professionals, though few discuss it openly and honestly in the public sphere.

Henry Marsh bucks the trend. In his 2017 memoir *Admissions: A Life in Brain Surgery*,[6] he vividly recalls his own petulant reaction when a nurse was instructed by a speech therapist to insert a nasogastric tube that Marsh—the surgeon—did not believe the patient needed:

> 'Take the tube out,' I said, between gritted teeth. 'It should never have been inserted in the first place.'
> 'I'm sorry Mr. Marsh,' the nurse replied politely, 'but I won't.'
> I was seized by a furious wave of anger.
> 'He doesn't need the tube!' I shouted. 'I will take responsibility. It is perfectly safe ...'
> 'I'm sorry Mr. Marsh,' the hapless nurse began again. Overcome with rage and almost completely out of control, I pushed my face in front of his, took his nose between my thumb and index finger and tweaked it angrily.
> 'I hate your guts,' I shouted, turning away, impotent, furious and defeated, to wash my hands at the nearest sink. We are supposed to clean our hands after touching patients, so I suppose the same applies to assaulting members of staff.

Again, we might suppose that such unsavory medical dynamics are a vestige of yesteryear. Sadly, interns, nurses, and other "lower" mortals in medicine still endure a courtly clinic hierarchy. In 2022, a systematic review of 25 studies of 30,000 surgical residents (doctors who have completed their medical school education but are still in training) found that 63 percent had experienced bullying.[7] Senior surgeons were the most common perpetrators, and of the victims,

71 percent did not formally report the behavior, of whom 51 percent said this was due to fear of retaliation. In the US, an Institute for Safe Medication Practices Survey found that 84 percent of nurses encountered doctors' reluctance or refusal to answer questions, or return calls, 71 percent experienced doctors' "condescending or demeaning comments or insults," and 21 percent had objects lobbed at them by doctors.[8]

It is not only bullying, but also sexual harassment. In the UK, findings from a study of 1,400 surgical staff found that a third had been sexually assaulted by a colleague in the previous five years.[9] In the US, a survey of more than 6,700 residents enrolled in surgery programs found that 80 percent of women experienced gender discrimination, with 43 percent of women saying they had experienced sexual harassment in the workplace.[10]

Bad behavior is not confined to surgical settings. In another US study of over 800 doctors, almost three in four had witnessed disruptive behavior during the previous month and, like Henry Marsh, more than a quarter confessed their own conduct had fallen short of acceptable standards at least once in their career.[11] In 2020, in a major NHS survey of 569,000 employees across 300 health organizations in England, one in five staff reported being bullied by colleagues.[12] Studies also show female staff or those who belong to a minority group are more likely to bear the brunt of bad behavior.[13,14]

Summing up the situation, Canadian GP Sally Mahood says: "We all see physicians deride nurses or other healthcare workers, 'dump' patients, violate confidentiality, ignore rules, use inappropriate language or demonstrate inability to work effectively with others."[15] As Henry Marsh's confessional example illustrated, top-tier doctors—individuals whose prestige begets power—are more liable to lash out than those lower in the hierarchy, especially if they perceive their status is being challenged or insulted.

Medicine is still waiting to embed the right solutions to effect culture change. Ultimately, in this standby mode, it is patients who pay the price, but there are other consequences too. Bullying takes a massive health toll on the immediate victims—health workers.[16,17] A recent NHS England report also estimated that bullying and harassment cost taxpayers £2.281 billion per annum in days lost due to employee sickness, reduced productivity, and industrial relations fees.[18]

Nagging, neglect, and unhappy nuptials

While human relationships in healthcare are not a paragon of domestic bliss, more hope is placed on the bond between doctors and technology. There is a fuzzy, romantic assumption that physicians and machines will somehow be compatible bedfellows.

Some of this optimism could be derived from historic events in the game of chess. Back in 1997, Garry Kasparov was the greatest chess player of all time. Then he played a match with IBM's Deep Blue supercomputer and lost. A few years later, in 2005, with greater advances in technology, including computer-processing power, a brand-new chess competition was launched. Known as the Freestyle Chess Tournament, hosted on the Playchess.com platform, it invited a wide variety of competitors: chess grandmasters, human–computer teams (known as centaurs), and computers playing independently.

Who or what won? It was not a grandmaster, nor was it AI. Two amateur players, Steven Cramton and Zackary Stephen, working with chess computers, including "Fritz, Shredder, Junior, and Chess Tiger," took the trophy as a team. The so-called "ZackS" team were a killer "centaur" combination that combined their strategic understanding of chess with the computational power of AI. The win neatly exemplifies the idea of extended cognition: that our thinking isn't confined to our brains but extends to the tools and environment around us.

Later, reflecting on this victory, Garry Kasparov concluded that "Weak human + machine + better process was superior to a strong computer alone and, more remarkably, superior to a strong human + machine + inferior process."¹⁹ The observation is now known as "Kasparov's Law."

Many leading digital health experts, including doctors, champion this view. For example, Dr. John Halamka, an emergency medicine doctor, digital health leader, and Dwight and Dian Diercks President of the Mayo Clinic Platform, says: "AI will never replace a competent physician. That said, there is little doubt that a competent physician who uses the tools that AI has to offer will soon *replace* the physician who ignores these tools."²⁰

Studies increasingly suggest that patients, too, are giving their blessing to the marriage of doctors and AI.²¹ Like many doctors, most US patients view AI as a complement to traditional patient care rather than as a substitute for it.²² A recent Pew Research survey found that around four in ten Americans would feel comfortable if their provider "relied on AI for their medical care."²³ Like Liam, when people are worried or when diseases are serious, some research suggests that they especially want AI to have a say.²⁴ In the UK, a survey conducted in 2024 found that more than half of the public supported the use of AI in the NHS for patient care, with one in three unsupportive.²⁵

This idea of a partnership between doctors and technology looks like a happy compromise. But the vision misses subtle lessons from the ZackS chess victory story—and also overlooks some surprising recent twists in chess's AI evolution.

First, in the medical vision, AI seems to be envisaged as a glamorous, subservient assistant to doctors who function as "head of the healthcare household." But recall that in the winning "centaur" team, the humans in the loop were amateurs. In the Freestyle Chess Tournament, weaker ("lower-status") players outperformed traditionally higher-status players when aided by AI. Perhaps, by demonstrating more humility and greater appreciation for the nature of the collaboration, they engaged with the machines in a particular way. Doctors, on the other hand, are more like grandmaster figures—they

THE SHOTGUN MARRIAGE OF MAN AND MACHINE

may be so knowledgeable that they too readily dismiss what the technology suggests.

The story becomes even clearer when we consider the long history of "man and machine" in medicine, which hasn't exactly been a heavenly match fit for Hallmark movies. Since the 1980s, digital tools in the form of "clinical decision support systems" (CDSS) have been incorporated into clinics. Consequently, a lot is known about the human–technological dynamic.

CDSS usually utilize patients' personal information, held in electronic health records, to offer suggestions—via reminders, alerts, and prompts—about what to do next. Many of these tools primarily rely on "expert systems" (which we met with Dr. Allen Wenner and medical history-taking). These algorithms can advise doctors on a range of issues; for example, that giving two drugs together—say, prescribing diphenhydramine (an antihistamine drug) with dimenhydrinate (used to treat motion sickness) might make a patient excessively drowsy. Some clinical decision support tools are designed to assist doctors to create a list of possible diagnoses, based on the patient's symptoms—what doctors call "differential diagnosis." Tools such as Isabel—which we'll explore later—require doctors to manually enter the symptoms into the tool, which then comes up with diagnostic probabilities and lists of relevant information.

Do any currently available tools elevate doctors' expertise? Certainly, there are a few success stories. A study at Cedars-Sinai Health System in Los Angeles, California found that doctors who followed CDSS alerts improved patient care.[26] Patients whose doctors disregarded alert notifications experienced a 6.2 percent increase in their hospital stay. When doctors overlooked the computer's advice the cost of patient care rose by 7.3 percent.

CDSS can also help to reduce delays in diagnosis, and improve accuracy.[27,28] One study found that primary care doctors who used these tools eliminated racial and ethnic disparities in patients' blood pressure control.[29] Researchers at Johns Hopkins School of Medicine also discovered that use of a computer aid to detect blood clots in the veins eradicated racial and gender disparities among hospitalized patients.[30]

In the main, however, the effectiveness of current CDSS is uncompelling. Aside from studies where doctors are "forced" to use these tools, evidence that they work in the real world of medicine is lacking.[28,31] For example, one review of more than a hundred studies into these tools concluded that they achieved only small improvements, with a 6 percent rise in the number of patients receiving the care they needed.[32]

The truth is that doctors override or ignore most clinical support alerts. For example, in 2014 an examination of over 2 million medication orders in outpatient settings prompted nearly 160,000 alerts with half them overridden.[33] Half of these overrides, in turn, were judged to be clinically important and ignoring the messages may even have harmed patients. Unheeded alerts included warnings about duplicate drug prescriptions, patient allergies, and dangerous interactions with other prescribed medications.

Some CDSS could even cause harm. Clinicians are more likely to ignore alerts when they're behind schedule—which is most of the time. Bombarded with annoying memos about what to do next, doctors report feeling browbeaten, and burned out from what they perceive as constant computer nagging. "Alert fatigue" is a phenomenon that is recognized by all physicians working with an electronic health record. This technostress is also connected with increased risk of errors,[34] with studies implicating CDSS as a cause of safety problems, especially medication mistakes.[35]

The phrase "right partner, wrong timing" describes the situation where humans and computers might be well-suited to work together but, due to circumstances, are unable to pursue or sustain a successful "relationship." Recall that chess grandmaster Garry Kasparov concluded after the ZackS team's success that a "better process" was key. In medicine the work environment forms part of the process in which these tools are used. Yet in noisy, overstressed settings doctors feel especially harassed by these alerts, and may end up ignoring them.

However, just as in life, "external" factors to relationships aren't everything either. Compatibility inevitably includes the qualities that both technology and humans bring to the table. Let's consider further evidence that so-called "strong humans"—doctors and other experts—may not in fact do well with machines.

Blowing hot and cold

Liam took what his Apple Watch "said" seriously. In contrast, Liam's cardiologist was contemptuous of it.

Was his cardiologist justified? Let's look at the evidence. Since 2018, the Apple Watch has been cleared by the Food and Drug Administration (FDA) for detecting irregular heartbeat, including AFib.[36] FDA clearance was intended for people aged 22 and older who, like Liam, have no history of AFib or any other heart problems. For these people, a wealth of research shows that the Apple Watch is highly accurate at identifying when AFib is present.[37,38] The watch is also very good at correctly identifying when AFib is not present, meaning it has a low rate of false positives.

Among people who already have AFib or irregular heartbeats, the Apple Watch isn't always as effective. In some cases, it might miss an AFib episode, leading to false negatives. For these patients, the watch works better as a supplementary tool, helping people to manage their condition alongside the help of health professionals.

This means Liam's presentation should have been a classic scenario when Apple Watch readings were taken extremely seriously by his cardiologist—but they weren't.

Dr. Jennifer Logg is a psychologist at Georgetown University in Washington DC who could have predicted both Liam's and his cardiologist's responses. Consulted by decision-makers in the US

Senate, Air Force, and Navy, Dr. Logg is a pioneer in understanding the relationship between humans and machines. She says that in general, laypeople—much like Liam—tend to readily defer to algorithms. Sitting down with me for a video call, she told me we often view machines as more neutral, less biased, more consistent, and smarter than other humans—a phenomenon Dr. Logg calls "algorithmic appreciation."[39]

While laypeople often openly defer to algorithms, her work also shows that experts are more frequently dismissive of them.[39] It took Dr. Logg two years to obtain access to US national security professionals. After being granted it, she set up an experiment where 70 of these experts were given the same information as a lay sample, together with additional sources of advice, and requested to make predictions. What she found was fascinating: the experts discounted algorithmic advice, hurting their accuracy. Dr. Logg told me, "The people off the street who took my online surveys were making more accurate geopolitical predictions than people who had between 5 and 30 years' experience."

Dr. Logg's studies of security experts are highly relevant to medicine. Her findings echo those of Paul Meehl, a clinical psychologist and philosopher of science who was one of the first to see cracks in clinician–computer relations. In the mid-twentieth century, Meehl examined the accuracy of clinical versus statistical methods in decision-making. Published in 1954, his seminal work, *Clinical versus Statistical Prediction: A Theoretical Analysis and a Review of the Evidence*, emphasized that statistical or algorithmic methods were often equal or superior to the clinical judgments made by experts, yet clinicians routinely ignored these outputs in favor of their own opinions.[40]

Today, substantial research supports the observation that experts can be unreasonably biased against algorithms, favoring their own (sometimes flawed) human decision-making instead.[41] This kind of apathy is referred to as "algorithmic aversion"—a phrase coined in 2015 by psychologist Dr. Berkeley Dietvorst of the University of Chicago.[42]

THE SHOTGUN MARRIAGE OF MAN AND MACHINE

Like Jennifer Logg, Berkeley Dietvorst's research shows that non-experts often display algorithmic appreciation. However, if increasingly exposed to algorithmic errors, humans—including non-experts—tend to lose confidence. When this happens, people exhibit higher thresholds of intolerance for algorithms than for fellow humans. For example, in one study that Dr. Dietvorst conducted, people were more likely to choose a human forecaster even when the human produced twice as many errors than the algorithm.[42]

Translating these findings to healthcare, investigations demonstrate that doctors blow both hot and cold toward algorithms. Dr. Olga Kostopoulou, a psychologist at Imperial College London, is one of the few health researchers thinking hard about algorithmic appreciation and aversion. For example, investigating the use of cancer risk tools, Dr. Kostopoulou found that general practitioners often either completely ignored or entirely accepted algorithmic advice.[43,44]

Or take medical imaging, where we already know AI has potential to be hugely disruptive. A study led by researchers at Harvard and MIT found that AI predictions were more accurate than almost two thirds of radiologists.[45] On average, when radiologists had access to AI predictions, this didn't strengthen their performance since they frequently undervalued the AI's input. Yet research shows that when radiologists actually engage with AI their performance can improve.[46,47] Making matters more complicated, when AI makes errors, and doctors unthinkingly rely on AI predictions, they are more likely to diminish their own accuracy too.[48]

Effective adoption of technology could improve diagnostic accuracy, disease risk estimates, and referral decisions. If an algorithm provides insights based on a comprehensive analysis of data, but a clinician only partially relies on the outputs, placing more emphasis on their own opinions, they are failing to give due credit to the algorithm. On the other hand, sometimes dismissing algorithms is exactly the right thing to do. Understanding when and why clinicians and patients blow hot and cold toward technology is essential. Equally important is determining how best to nurture this relationship—if it should exist at all—always with the patients' best interests to the fore.

To better explore the intricacies of this entanglement, we'll next invite clinicians and computers onto the therapist's couch. Whatever is going on, one thing is clear: we need a firmer understanding of the psychology of clinician–algorithm interactions. And, as in any mediation, the outcome requires both technological and medical communities to acknowledge that there is a serious problem to be solved.

Clinicians and computers on the couch

"It's partly the arrogance of the doctors. It plays a huge role in this. They come out of medical school arrogant."

British entrepreneur Jason Maude is both damning and matter-of-fact about doctors' resistance to algorithmic advice. As CEO and co-founder of Isabel since 1999, he knows the profession intimately: his is one of the best-known CDSS on the medical market. Given a range of symptoms, it works by generating a list of potential diagnoses. The origins and development of the tool are profoundly personal to Jason—Isabel is the name of his daughter, and she is the reason the tool exists.

In 1999 at the age of three Isabel caught chickenpox. She soon became seriously ill, with purple blisters spreading across her stomach. The Maudes rushed their daughter to the emergency department, but doctors were uncertain about what was wrong. Only on referral to St Mary's Hospital in Paddington in London did doctors correctly diagnose "necrotizing fasciitis"—a flesh-eating bug—and operated immediately. After eight weeks in hospital, Isabel returned home. Shaken by the experience, Jason, a former stockbroker, vowed to make medicine safer. He switched trades, devoting his career to the other "Isabel" in his life—the development of an online symptom checker for doctors and patients.

"The technology side has been easy," he tells me. "The hard part is how you get these tools adopted. I have struggled over many years to figure out why that is. Doctors see it as a threat.... We've had people say, 'I think when I use this, I'm cheating.'"

Jason adds, "If I bought a sat nav for the car, I didn't struggle over when I would use it. It was pretty obvious. And you think, 'Why is this so difficult? Why is this so difficult for you to get?'"

Echoing this, surveys show doctors fear that AI will encroach on their professional autonomy, with potential for them to "lose control" of the consultation.[49] Studies also show doctors often fail to use, or see the need for, computer support.[50,51]

What Jason Maude perceived as arrogance might be partly explained by over-confidence and doctors being excessively certain of their own judgments. Because the "art of medicine" is a dark art, doctors are often wrong when they think they are right, and vice versa. This means they may not be "listening" to all that the AI is "telling" them. Sure enough, even when "forced" to use Isabel—such as happened in a clinical study—doctors don't strengthen their performance. A multicenter study of 1,200 patients across four Swiss emergency rooms put the diagnostic tool Isabel to the test.[52] Fifty percent of patients were randomized between doctors using Isabel and those relying on their usual methods. The verdict? Isabel made no difference—diagnostic accuracy remained unchanged.

Even when used, tools like Isabel may come too late in the diagnostic process and reinforce what doctors already think, strengthening their existing biases instead of challenging them. One explanation is that, since they form their diagnostic impressions so rapidly, cognitive biases influence how doctors use computer aids—including whether they use them at all. For example, an Australian study of a cancer risk tool among GPs found that medical experience and convictions about clinical intuitions were key determinants of tool use.[49]

Other support for this comes from the work of Dr. Olga Kostopoulou of Imperial College London. She led a study that found doctors' data entry in electronic health records was patchy and incomplete, reflecting their early diagnostic impressions.[53] Doctors

DIAGNOSIS AND TREATMENT

typically recorded data pertinent to their presumed diagnosis of the patient, rather than documenting a comprehensive account of the patient's symptoms. She also found that clinical information searches tend to be adopted in ways that preserve doctors' initial judgments rather than challenge or augment them.[54]

We've hinted that doctors might be viewing tools like Isabel as akin to a person. We've met this human attribute before: people anthropomorphize artifacts, systems, and even institutions. Before we get to how clinical decision support tools may—or may not—evoke anthropomorphism and what this might mean for medical decision-making, it is helpful to understand more about why and how this effect arises. Throughout this book we've observed that evolution equipped our minds with highly specific capacities—what cognitive scientists call "modules"—selected for by natural selection to solve all kinds of recurrent problems in our hunter-gatherer environments and to process information in a very specific manner.[55,56]

While experts debate how modular the mind really is, they agree that our brains come with a lot of built-in skills and capacities. These range from language learning to recognizing danger, from spotting cheaters to knowing that objects fall when dropped. No one had to teach us these things in preschool. For example, consider a common challenge our ancestors faced in the Stone Age—spotting deadly predators like snakes. Those who quickly noticed signs of snakes, such as their sleek looks and smooth movements, had a better chance of surviving by reacting fast.[57] This ability to detect snakes, however, isn't perfect: even a piece of twisted wood might make us jump. It's safer to mistakenly react to a false alarm than to ignore a real threat. In other words, we can be triggered by other things, not just the real danger our mindware evolved to detect.[56]

Similarly, the design of technology can sometimes trigger our brains to perceive devices as if they were human. Is there any evidence that doctors do the same? That is, do they treat and perceive the AI they use as human? Findings are indirect but tantalizing. For instance, when CDSS generate frequent alerts, doctors often perceive these as the machine "nagging" or even "scolding" them, like an overbearing person who constantly gives unsolicited advice.

Liam's cardiologist was so piqued that the Apple Watch was invited into the consulting room that she refused to follow up with a proper examination. Earlier we observed how the psychology of status can be activated during interactions between doctors, nurses, and trainee staff. Doctors may not bully or harass AI, but algorithms might engage a sense among doctors that there is another person—an upstart or rival—squaring up to them and their expertise. This might help to explain why physicians tend to ignore, or even get angry with, algorithms.[58,59] Other explanations are possible too. Notice also that Isabel—the clinical decision support tool—offers doctors a list of possible diagnoses, which are also often ignored by physicians. Perhaps too many possible options signal lack of confidence and therefore incompetence to doctors.

Alternatively, researchers might discover that doctors and patients don't anthropomorphize these tools at all. In this scenario, humans might simply perceive algorithms as what they are—technology that is designed for specific jobs. In some cases, we might consider them to be inferior pieces of kit. To explore this scenario, consider the reflections of the late American philosopher and cognitive scientist Daniel Dennett.

Dennett described anthropomorphism as invoking what he called "the intentional stance."[60] This means interpreting the actions of entities, animals, or complex systems as though they possess beliefs, desires, and intentions. Additionally, Dennett suggested that humans often naturally adopt a "design stance"—an approach where we interpret and predict the behavior of objects and systems by assuming they are *designed* with a purpose.

An example of Dennett's "design stance" is how we might understand the behavior of a thermostat. When we look at a thermostat, we don't consider the intricate details of its internal workings. Nor do we treat it as if it were a thinking entity with beliefs or desires. Instead, we interpret its effects based on its designed purpose: to maintain a room's temperature at a set point. We understand that if the temperature falls below this set point, the thermostat is supposed to turn the heating system on. If it exceeds the set temperature, its job is to turn the heating off. When we adopt the design stance, we straightforwardly predict and understand the thermostat's behavior based on its function.

It is quite possible that people simply adopt the design stance with machines. Again, although we can't offer definitive answers, there are hints this sometimes happens. Recall that Dr. Berkeley Dietvorst's research shows that people are more likely to abandon algorithms if they make a mistake, losing confidence in them faster, and coming down harder on them.[42] This almost perfectionist perspective on algorithms might arise from the view that machines are instruments designed for a task. Perhaps, seeing algorithms fail, even once, leads humans to consider it a bit of broken kit—rather like a thermostat that malfunctions. In contrast, we tend to forgive humans because, unlike rigid tools, we recognize that people are more flexible and can learn from their mistakes.

Adding to the complexity, people can adopt both the design and intentional stances, often switching between the two, depending on the context and their perceptions of the system or object at hand. For instance, a doctor using a clinical decision support tool might primarily adopt the design stance to understand how the system processes information, but instinctively adopt the intentional stance when the system makes certain recommendations.

Our ability to fluidly switch between different interpretative stances—perhaps in this case from design to intentional stances—will depend on the cues and attributes of the algorithms. Supporting this, studies also show that people's perceptions of algorithms change depending on what they see them do—the design, reliability, and ease of use of technology. For example, in one study patients were more reluctant to use AI if they believed that it would not understand their unique needs as well as human clinicians would; resistance decreased when the AI was presented as personalized.[61] Another experimental study entitled "Watch me improve" found that when people perceived algorithms as learning from their mistakes, they trusted them more.[62] It seemed that they shifted from the design to the intentional stance, perceiving the machines as capable of acquiring knowledge, and having thoughts and beliefs just like people.

Jason Maude worried that doctors were being arrogant in rejecting traditional decision support tools. But, as we've seen, resistance to technology is complex. From the environment doctors use them in to their perceptions about the digital tools themselves, and how these tools relate to their own expertise—all these factors matter.

We've thrust doctors' interactions with technology onto the couch for a therapy session that has been brief, raising more questions than it answers. Complicating matters further, newer, more captivating digital tools are upping the ante.

Tying the knot for tomorrow?

We've met them before—enter the Digital Don Juan, the Silicon Seductress: AI powered by large language models (LLMs). Compared to their AI ancestors, ChatGPT, Gemini, and DeepSeek seem like they've been schooled in charm. More of us are using these bots in our jobs and daily life, and doctors are using them increasingly too.[63] Commercial chatbots like ChatGPT risk patient privacy and safety; yet so widespread is doctors' interest that some medical organizations were compelled to issue warnings and advisories about using them in clinical practice. For example, in our 2025 survey of 1,000 UK GPs, one in four reported using tools like ChatGPT for clinical tasks—27 percent for differential diagnosis and 24 percent for suggesting treatment options. GPs were willingly adopting these tools without being "forced" to do so: without any organizational prompts, guidance, or workplace training.

DIAGNOSIS AND TREATMENT

What makes these tools so appealing? Unlike traditional decision support tools, they don't "hector" with unsolicited advice, are more "confident" in their interactions, and seem to afford a supportive, consultative role. The back-and-forth, with repeated rounds of "dialogue," is suggestive of fluency, courtesy, and attentiveness. LLM bots can even change their "mind," something that strongly induces us to treat these tools like people. Studies show that the more we perceive chatbots as having thoughts, beliefs, and desires, the more moral responsibility we assign them.[64] These apparently "personal" qualities can carry huge consequences for clinical decision support. If patients pour their hearts out to machines—given the right interface and design—doctors may well do the same in professional settings.

When it comes to the future of doctor–AI partnerships, emerging studies are proving hard to ignore—especially for patients. An experiment at Google Research and Google DeepMind recruited 20 medical professionals to look at 302 complex, real-life medical cases from *The New England Journal of Medicine*.[65] All doctors offered an initial list of differential diagnosis—those most likely conditions given the presenting symptoms. Afterwards, each case was reviewed by two doctors who were randomized into one of two groups: one group used search engines and standard medical resources to help them, while the other group used these tools plus ChatGPT-4.

Doctors who used ChatGPT-4 were more likely to have the actual diagnosis in their top ten list of possibilities: they achieved 52 percent accuracy compared with those who only used search tools, who scored an accuracy rate of 44 percent. Lacking any assistance whatsoever, doctors' initial accuracy was a meagre 36 percent. This seems to suggest doctors were indeed "listening" to what these chatbots have to say.

However, things are not as rosy in this relationship as they seem. Doctors seemed to dismiss the AI too often, perhaps treating it as an unworthy subordinate or viewing it as a piece of broken kit. We know this because ChatGPT-4 working solo scored 59 percent. In other words, doctors in the loop actually hurt the bot's diagnostic accuracy. Other studies support this finding. A well-publicized study published in *JAMA Network Open* in 2024 found that when doctors used ChatGPT-4 their diagnostic accuracy was 76 percent, and stood at 74 percent when relying on conventional sources.[66] However, proving that team sport isn't everything, the bot working alone excelled, achieving a 90 percent accuracy rate.

Medical-grade chatbots—LLM tools specially designed for healthcare—could be even more impressive.[67] Take a study of 1,066 consumer medical questions where doctors rated the AI model Med-PaLM-2 as superior to their peers on eight of nine measures, including accuracy, reasoning, and safety.[68] Impressively, Med-PaLM-2 also excelled in addressing bias and minimizing harmful advice, showcasing its potential to enhance clinical decision-making.

For those toying with the notion of a digital divorce, or holding out hope for a seamless marriage between doctors and AI, I offer some advice. First, it's fair to say that LLM tools haven't yet swept clinicians off their feet. AI is undoubtedly impressive, but, unlike the clear-cut outcomes of AI chess matches, we must be mindful that many of these medical tests and studies involve artificial set-ups. The real-world performance of AI in healthcare remains uncertain and demands closer evaluation.

Attesting to this, a systematic review of 519 studies published between 2022 and early 2024 revealed striking gaps in how large language models are examined.[69] Only 5 percent of studies used real patient care data to assess LLMs, and the vast majority focused on answering medical exam questions (45 percent) rather than practical tasks like diagnosis. Critical issues like fairness and bias were addressed in just 16 percent of studies. Clearly there is an urgent need for a comprehensive, practical evaluation of AI's safety and reliability.

Second, some argue that doctors might better learn to "talk" to AI, smoothing out the relationship. But as the two amateurs in the

DIAGNOSIS AND TREATMENT

"ZackS" chess team demonstrated with their devastatingly brilliant AI collaboration, we must also figure out the right human coworkers for medicine. Can so-called lower-level clinicians—nurse practitioners or perhaps a new breed of coworkers—strike the perfect balance? Doctors, as we've observed, face an existential challenge in the age of AI, fearing these tools as direct competition. Armed with just enough medical knowledge and a keen awareness of both their own and AI's limitations, like the ZackS team, new kinds of human workers might just form the winning partnerships healthcare has been waiting for.

Yet even this could be a medical mirage. The chess AI story takes another intriguing turn when we consider developments that go beyond Kasparov's Law. In 2017 an AI called AlphaZero beat all its competitors including human-plus-AI teams. This effectively ended the era of centaur dominance in chess. Today, no human or human–computer combination can rival top AI chess engines. Nowadays, chess AI operates with speed, precision, and creativity that even the best human configurations cannot match.

The lessons for medicine are obvious. While caution is essential, we've seen signs that AI might flourish away from the cramped constraints of clinicians' input. More to the point, technological advances are not abating.

OpenAI's o1 is an advanced AI system designed to perform complex reasoning tasks across various fields. Dr. Adam Rodman and a team of collaborators tested the o1 model, revealing its unprecedented AI ability to perform complex physician-level reasoning.[70] The AI far surpassed both ChatGPT-4 and human doctors. In differential diagnosis, o1 identified the correct diagnosis in 89 percent of cases, compared to ChatGPT-4's 73 percent. When managing intricate clinical decisions—the choices and actions taken to treat or address conditions—it scored a remarkable 86 percent accuracy—leaving ChatGPT-4 at 42 percent and doctors (even with AI support) at 41 percent. Even if we must proceed with vigilance, the rapid pace of AI advancement hints at transformative shifts in clinical decision-making.

Doubtless, by the time you read these words, the field will be full of similar studies and critiques examining OpenAI o1 and o3, and a variety of AI successors. At the time of writing, OpenAI's latest model, o3, has introduced advanced reasoning capabilities. One of the key tests it has been evaluated on is "ARC-AGI," a benchmark designed to see how well an AI can learn new skills it hasn't been specifically trained for—something closer to humanlike adaptability. On the high compute setting (where it spends more time "thinking"), o3 scored an impressive 87.5 percent, a significant improvement over its predecessor, o1.[71] More notably, it surpassed the average human performance of 80 percent. Again, how well these tools fare in the real world—and how they compare with doctors—will need to be fully explored.

In the long term, not everyone may mourn the dissolution of the relationship between doctors and digital tools. If doctors still need to feel in command of technology, to the detriment of diagnostic accuracy, we might wonder whether marriage is mutually enriching, or a match made in hell and worth annulling.

I asked Liam for his opinion. "I still want a doctor to sign off on my diagnosis. But if I had to choose between two doctors, and one was using AI, I would trust her more. I'd respect her more too because even doctors have their limits."

Although we're not there yet, the day may come when seeing a human doctor instead of Dr. Bot feels as risky as being a pawn on a grandmaster's chessboard. When that happens, the balance of power may shift away from doctors to patients and technology.

PART IV
EMPATHY

10

Doctors Getting Deep

Dr. Emir Ayan—not his real name—is a primary care physician in Boston, Massachusetts. I met him outside a coffee shop in the Longwood Medical Area, one of the most densely packed medical centers on the planet. This 213-acre hub includes the world's top teaching hospitals.

His shock of black hair swept back, Dr. Ayan resembles a debonair, golden-age movie star. Adorned in a custom-embroidered white coat, he physically personifies a hero healer.

Between gulps of Diet Coke and mouthfuls of Snickers, Dr. Ayan talks passionately about his job: "Family medicine was always my ambition. A lot of doctors look down on it. But I'm a people person, and the human side of practice—the variety of patients—is the reason I picked it. My mom was a primary care doc too, and her father before her, so in some ways, it's the family business."

Today is like any other at the Longwood practice where he's worked for nearly 15 years. Dr. Ayan will see around 20 patients—roughly 100 per week. Compared with other primary care doctors in the US, his workload is average. It also compares favorably with doctors in other countries. In the UK GPs can see between 30 to 40 patients per day. In China doctors might see a staggering 80 patients daily; and in Bangladesh, due to fewer doctors and higher patient demand, medical visits last an average of 48 seconds. From an international perspective, Dr. Ayan has a relatively boutique figure of around 2,300 people allocated to his care.

His daily tasks are diverse: from performing routine physicals to discussing test results; from treating minor maladies to delivering life-changing news; and from making worrisome referrals to reassuring the worried well. Around half of Dr. Ayan's day is spent on paperwork. During and between visits, data entry vies with patients for his attention.

Today we discuss bedside manner: "I'm always conscious about providing respectful, empathetic care. Whatever the problem, my job is to recognize that every person sitting in front of me has unique needs. They need to see that I care." Over his career, Dr. Ayan says he has spilled tears over patients, some of whom he will never forget.

"A few years ago, a young guy really opened up to me. He'd suffered severe trauma and abuse as a kid. It was the first time he'd spoken to a clinician about it. His case affected me deeply. Maybe in another life, he was the kind of guy who could have been a high school friend."

"More recently—last fall—there was a case of an older lady with breast cancer. I knew her pretty well over the years—she had a wicked sense of humor and reminded me of my grandmother. This patient came to see me too late—the cancer was inoperable. If only she hadn't delayed. All I could do was arrange palliative care. For a few weeks it really played on me."

Exhaling deeply, then draining the last of his Diet Coke, Dr. Ayan adds, "It's all part of the job—the good stuff and the bad. In primary care we see it all." Conscious of the people who need him, Dr. Emir Ayan darts back inside the clinic.

Doctors have not always worn their hearts on their crisp white sleeves. And medicine has not always encouraged warm, fuzzy feelings. Canadian physician William Osler (1849–1919)—often called the "Father of Modern Medicine" and credited with putting the profession on a scientific footing—believed the doctor's temperament should be one of detached concern. Osler advocated the virtue of calm, critical distance, with doctors upholding a "coolness and presence of mind" and "impassiveness" in patient–physician interactions.[1]

Contemporary healthcare is more effusive about empathy. In 1993, the World Health Organization declared it "the heart" of doctor–patient communication.[2] In his highly acclaimed book, *Deep Medicine: How Artificial Intelligence Can Make Healthcare Human Again*, cardiologist and AI champion Dr. Eric Topol enthused, "Empathy is the backbone of the relationship with patients."[3] Medical educators agree. Empathy is now listed as an essential learning objective by the American Association for Medical Colleges, with increasing calls to test medical school applicants on their empathetic aptitude.

In this chapter, slicing through empathy's fluffy epidermis, we'll discover that doctors being empathetic is a far more complex idea than is often recognized. Doctors struggle to practice it consistently—and, paradoxically, physician empathy can even undermine our care.

Physician, know thyself

Like priests and actresses, many doctors describe their career as a calling. In 2018 a survey of 2,400 US students conducted by Medscape, a major news outlet for American doctors, reported that "the desire to help those in need" was the leading reason for choosing a career in medicine—cited by around 85 percent of trainees.[4] There is no doubt that other motivations contribute too, including high standing in society, a good salary, or simply following in family members' footsteps, but for most doctors, there is a genuine vocational tug.

It is perhaps partly for this reason that physicians sometimes struggle to accept that their capacity for empathy can go AWOL. This was a revelation brought home to me when debating the topic with an esteemed GP during a medical-literary festival in Cheltenham in the UK. At the hint of doubt, the individual thundered: "I am empathetic toward ALL my patients!" Such defensiveness, as we'll shortly discover, is wholly understandable even if the doctor's declaration is likely to be untrue.

Yet it is also fair to say that physicians genuinely do experience intense joy and—like Dr. Emir Ayan—can experience real grief for patients too.

In his elegant, bestselling memoir, *Do No Harm*, the British brain surgeon Henry Marsh reflected on one such distressing occasion.[5] A particularly poignant passage describes Marsh caring for David H., a patient with a low-grade astrocytoma, a slow-growing brain tumor. At their first meeting, Marsh recalled an energetic, fit young man in his early thirties. A successful management consultant who was married with a young family, David "was a person of great charm and determination, who managed to turn everything into a joke."

Over a period of 12 years, he met David on and off, operating on him three times. However, there came a point when the cancer was inoperable. After disclosing the news to David, his wife rose to thank their surgeon. "I buried my face in her shoulder, holding her fiercely for a few seconds." Leaving the bedside, another doctor approached him in the corridor but, Marsh writes, "I waved my hands despairingly in the air and walked away, imitating the staggering walk of a drunk, drunk on too much emotion."

Searing memories—such as this one—give us valuable insight into the experiences of practicing doctors. But to fully investigate the quality of care, it is vital to go beyond unique cases and episodic memories. More generally, studies in cancer care show that oncologists underrate the level of distress experienced by patients, and overrate their own bedside manner.[6,7] Nor are clinicians the best judges of how consistently and how well they deliver empathy.

The road to enlightenment

For many doctors there is a lightbulb moment when they finally do witness care from the other end of the stethoscope. The bestselling

DOCTORS GETTING DEEP

medical memoir *In Shock: How Nearly Dying Made Me a Better Intensive Care Doctor*, by Dr. Rana Awdish, a critical care doctor, describes the author's experiences of sudden life-threatening illness, and a very personal reckoning. The book opens with Awdish heavily pregnant and in extreme pain. Rushed to hospital, she describes what happens when a junior doctor wheels a portable ultrasound machine to her bedside.

> "Bear with me," the obstetrical resident warned, his foot tangling with the cord. "I'm not great at these yet." He didn't need to be. Here I could still be a physician. From the first grainy images I could see the small ventricles still and pulseless.... "There's no heartbeat." The words cascaded out of me on a torrent of agonized breath.
> "Can you show me where you see that?" he asked....
> As his question echoed, I discerned a tone in his voice that I hadn't initially noticed. It was genuine curiosity. I realized, with an uncomfortable tug of recognition, I was indeed, not a person to him, but a case. And an interesting case at that.... I affixed my eyes onto his, willing him to see me.[8]

Rana Awdish's miscarriage is the start of a prolonged period of critical illness, multiple hospital admissions, and near-death experiences. Her dramatic story culminates in an encounter with a doctor whose behavior, Awdish writes, was the last straw: "I was done. I was done tolerating patronizing physicians. I was done being made to feel as if I should accept the person whom my life depended on."

Clinician-cum-patient memoirs make for particularly compelling reading, and this ever-growing sub-genre typically depicts humbled doctors undergoing Damascene clinical conversions.

Take Professor Ashish Jha, a physician and academic, who served as the White House COVID-19 response coordinator from 2022 to 2023. Writing in *KevinMD*, an online blog for a mainly medical readership, Jha recalls his own epiphany after rollerblading with his kids, suffering a particularly nasty fall, and dislocating his shoulder. Lying in agony on the path and waiting for help, Jha is struck by the kindness of passersby eager to assist.

Arriving at the local hospital—and now among his fellow clinicians—he witnessed a stark contrast. "The biggest lesson for me," observed Jha, "was this was not an extraordinary story at all.... [W]hen we walk into a hospital, we give up on being people and become patients. We stop receiving care, the way I did on the bike path. Instead, we receive services. And when you are in pain, the difference between care and services is stark."[9]

Or consider the words of neurosurgeon Dr. Paul Kalanithi, whose posthumously published memoir on medicine and stage IV metastatic lung cancer, *When Breath Becomes Air*, is now a deserved classic. Kalanithi put it thus: "How little do doctors understand the hells through which we put patients."[10] Crucially, medical memoirists arrive at the same revelation as many other patients: healthcare often lacks humanity.

Robotic doctors

Doctors are not deliberately obtuse, or intentionally cold-blooded. Nonetheless, the experiences of Drs. Awdish, Jha, and Kalanithi are not freak incidents. A systematic review of patient surveys concluded that "rankings of practitioner empathy are highly variable."[11] Indeed, deficits of empathy in healthcare are the subject of news stories across the world.

In the UK, for example, in the early 2000s, an investigation was launched into the treatment of patients at Stafford Hospital, where concerns were raised among patients and their family members. A subsequent public inquiry led by Robert Francis, QC, reviewed more than a million pages of evidence, incorporating testimonies from 290 witnesses. Published in 2013, among its key findings the report identified "lack of consideration for patients," "lack of openness to criticism," "defensiveness," and "failure to put the patient first in everything that is done."[12]

Most instances of breakdowns in bedside manner are less dramatic. Auditory and visual recordings provide a valuable way to analyze the quality of medical interactions. This body of work demonstrates that doctors can be highly compassionate. Yet, even when aware they are on tape, they can also be surprisingly robotic; one study diplomatically concluded, "Doctors exhibit minimal receptiveness to patients' lifeworld disclosures," focusing instead on "'textbook' ... biomedical agendas."[13]

Delivering serious diagnoses might be considered the ultimate test of empathetic care and, in the US and Europe, cancer is ranked as the most feared illness. Starting in the 1990s, health researchers began to probe patients' first-hand experiences of their cancer care.[14,15] In 1996, investigators carried out over a hundred audiotaped "bad news cancer consultations" at a large London teaching hospital.[14] Investigators found doctors disclosed 4.5 times more biomedical information than social or emotional talk: "levels of psychological probing and counselling from clinicians were very low. ... Some patients did disclose their psychosocial concerns, but in the majority of cases these were not pursued by clinicians who kept to a rigid biomedical agenda." A few years later, in 2002, researchers in Australia replicated the study with nearly 300 patients, reaching a similar conclusion:[16] "doctors effectively identify and respond to the majority of informational [biomedical] cues; however, they are less observant of and able to address cues for emotional support." More recently, in 2016 another study of 46 newly diagnosed breast cancer patients highlighted the "low emotional support from surgeons."[17]

Pediatric intensive care conferences—when doctors meet with the parents of hospitalized children—are often deeply distressing events. In 2018, a study of nearly 70 such conferences reported that doctors did a good job of recognizing family members' emotions but signals of empathy were erratic.[18] Pediatricians responded to parents' emotional cues only a quarter of the time; about a third of the time they simply replied with more medical talk. Investigators dubbed these "missed opportunities" to comfort distressed parents. For example:

> *Family member*: I told them when he cries so loud and his whole body turns red he gets very still—it really worries me. It wasn't just the cry, it was his body language.... It worried me.
> *Physician*: The most common reason for a baby to do that is actually reflux.

Doubtless, the worried parent was left feeling more anxious and confused.

Adult primary care is also the scene of Dr. Spock-style rejoinders. Although no such study has been done, it seems possible that many doctors might fail the Turing Test, which checks if a computer can talk like a human so well that people can't tell if it's a machine.

Take a US study which found that in 79 percent of appointments doctors "missed opportunities to adequately acknowledge patients' feelings."[19] Squandered chances included off-hand ripostes such as inappropriate humor or terminating the topic of conversation. Failing to secure suitably humane responses, around half the time patients brought up emotional issues a second or even a third time. Lack of empathetic responses even prolonged visits.[19] Again, since all parties knew they were being recorded, evidence of clinician coolness might be the tip of the apathy iceberg.

Off-putting bedside manner may also undermine patient testimony. One study provided a rare window into doctors' visits with elderly patients who were suffering from severe depression.[20] Transcripts revealed physicians generally adopted a caring demeanor. But sounding like Siri or Alexa gone awry, conversations were often marked by "awkward transitions." Android-esque responses included what investigators called the "chitchat" maneuver whereby talk of suicide short-circuited any sense of social appropriateness, with doctors diving into "rambling conversation." For example, one patient's desire to kill himself was met with the response, "Have you been over to CiCi's [an American buffet chain] lately?" Never mind suicide, have you tried CiCi's new pepperoni pizza?

You've lost that lovin' feelin'

Wonky bedside manner is often blamed on failures of medical education. Endorsing this perspective, during the three-year-long Francis Inquiry in the UK, the erstwhile Prince of Wales—now King Charles III—dispatched his own epistle on healthcare's failings.

"Are we doing enough to ensure there is sufficient empathy and compassion instilled throughout training in medical schools and in later hospital training?" he asked. Writing in 2012 in the *Journal of the Royal Society of Medicine*, the now king emphasized "the need to restore urgently a climate of care and compassion at the heart of our health services," arguing "there is much more that can be done to foster and enhance those age-old qualities of human kindness and compassion."[21]

In Britain, most doctors were probably too busy with the day job to pore over the future monarch's missive. His views, however, are not fringe opinions. Doctors and medical researchers frequently point out that empathy dips at exactly the moment students meet their first patients. Like the British monarch, they commonly infer that medical school is where the rot sets in.

For example, according to British physician Rachel Clarke in her book *Dear Life: A Doctor's Story of Love and Loss*, "Empathy ... was under assault from day one of medical school. With biochemistry and anatomy filling our days, one way or another, people were reduced—either to chemical interactions or to corpses on slabs."[22] Emphasis only on cold science, it is assumed, starves students of the human side of care, rendering doctors emotionally anemic.

Since education is perceived to be the culprit, it is also viewed as the remedy: a different curriculum, it is claimed, could halt the hemorrhaging of empathy. In his latest book, *Deep Medicine*, Dr. Eric Topol proposed: "We need to rewire the minds of medical students so they are human oriented rather than disease oriented."[3]

Advancing the same theme, one of medicine's foremost weapons in the battle for empathy is a pedagogy called "narrative medicine," the brainchild of Rita Charon, a general internist and professor at Columbia Medical School in New York, who argues that medical practice should be structured around the narratives of patients. Dr. Charon claims, it is not enough to *understand* what the patient is going through; physicians should *feel* something of the patient's emotions too.[23]

This view enjoys strong support among the clinical cognoscenti at many medical schools. For example, Dr. Jodi Halpern, a psychiatrist and professor of bioethics and medical humanities at the University of California, Berkeley, whose book *From Detached Concern to Empathy: Humanizing Medical Practice* was declared a seminal work by the *Journal of the American Medical Association* (*JAMA*), urges that "empathy requires experiential not just theoretical knowing," and "The physician experiences emotional shifts while listening to the patient."[24]

From California's Keck School of Medicine to King's College London, narrative medicine is embedded into training to promote empathetic care. Attesting to mainstream support, *JAMA* specifically curates a special category of articles on narrative medicine alongside articles in more traditional medical fields such as oncology, neurology, and surgery.

The medical community's motivation to improve their bedside manner via education is wholly commendable. However, like a direct-to-consumer drug commercial showing people frolicking on beaches, the promotion of empathy in medicine oversimplifies the challenges. None of this is to deny the need for improvements in bedside manner. Rather, as we'll now discover, doctors are not optimally equipped to provide us with empathy. In fact, they may be the least well equipped of all people.

Making matters more complicated, demanding that doctors deliver empathy is not as desirable as we might suppose. Unlike the high-velocity voiceovers warning viewers about drug side-effects, we will also dwell on the fine print—the caveats omitted from the glossy, sometimes gooey, promotion of physician empathy.

The multitasking myth

King Charles blamed education for doctors' lack of sensitivity. But this is a medical misdiagnosis. Committing the logical fallacy "after this, therefore because of this," it is supposed that shortcomings in doctors' education are the only, or the most important, explanation for breakdowns in bedside manner. By embracing this idea, Charles and the medical community implicitly assume doctors can do it all, and moreover, can do it all at once: be genius diagnosticians, empathetic carers, and administrative whizzes.

Even leaving aside doctors' detested clerical duties, the dual demands of diagnostic brainpower and empathetic big-heartedness are considered fundamental to physicians' identity (though the first, more so). This twin aspiration is well captured in a published volume of student essays entitled *The Soul of a Doctor: Harvard Medical Students Face Life and Death*, where, in the foreword, a distinguished Harvard Medical School professor emphasizes that students must acquire what he calls "clinical binocularity," described as "the ability to think about people, and to relate to them both as individuals and in terms of the mechanisms that produce disease."[25]

Clinical binocularity sounds like an impressive medical skill, and from a patient perspective it sounds desirable to boot. It is, however, a physician fantasy. Psychologists Kevin Madore and Anthony Wagner of Stanford University have explored the problems with multitasking, and their work supports the conclusion that humans don't multitask at all:[26]

> The scientific study of multitasking over the past few decades has revealed important principles about the operations, and processing limitations, of our minds and brains. One critical finding to emerge is that we inflate our perceived ability to multitask: there is little correlation with our actual ability. In fact, multitasking is almost always a misnomer, as the human mind and brain lack the architecture to perform two or more tasks simultaneously.

Instead, they argue, people task-switch. While we might think we're multitasking when we write an email while talking to someone on the phone, what is really happening is: we write the email, then we talk, then we write the email—toggling back and forth. Yet during our visit the doctor is required to undertake multiple tasks: expertly gather and document our clinical information, perform examinations, undertake diagnoses, and communicate with patients in an approachable, understandable, and empathetic manner.

Recall how Dr. Rana Awdish willed the obstetric trainee to switch from deciphering an image on a screen to seeing a mother who had just miscarried. While the balancing act is tougher for trainees, experienced doctors struggle with task-switching too and, in moments of candor, they admit to the difficulties of juggling duties. Awdish honestly says, "listening is harder than it seems," and physicians "listen imperfectly, through a fog of ghosts and competing priorities."[8] Just as texting while driving is a bad idea, toggling between activities in medicine creates inefficiencies, and can ramp up the risk of errors.

Rather than advocating for clinical binocularity, Harvard Medical School social scientist and physician Omar Sultan Haque and psychologist Adam Waytz of Northwestern University highlight the practical dilemma this approach poses.[27] Medical multitasking, they argue, begets cross-eyed clinicians. According to Haque and Waytz, when doctors focus their attention on abstract thought, this automatically dials down the capacity for empathy; research supports this, revealing that empathy represses reasoning about abstract tasks. Neuroscience imaging shows that regions of the brain engaged with social reasoning, such as understanding other people's mental states, become deactivated when we engage in mechanical reasoning of the sort involved in solving science problems, and vice versa.[28] In other words: trying to be empathetic while thinking hard could undermine our doctors' diagnostic judgments.

Doctors need a splinter of ice in their hearts if they are to perform their clinical duties effectively. This insight is confirmed by doctors' keenest personal reflections. Writing about her early career, Dr. Rachel Clarke reflects,

> I'd moved on from my early days of struggling ineptly to site cannulas in my patients' veins, their grimaces at my clumsy efforts making me hot and sweaty with guilt. I'd learned that treating the task as a technical challenge rather than a human encounter gave the patient the best chance of a pain-free cannula. Paradoxically, being cruel was the best way to be kind. I would talk gently and soothingly as I decisively stabbed them, but inside I was ruthlessly focused, feeling nothing at all.[29]

Haque and Waytz refuse to sugarcoat the trade-off: "The problem-solving benefit of dehumanization," they argue, "... may be especially important when the pressure to deliver efficient and effective care is high."[30] In other words, if we observe a doctor doing it all—displaying the diagnostic brilliance of Dr. Gregory House with the bells-and-whistles empathy of Dr. Quinn (*Medicine Woman*)—there is a fair chance they're a soap character.

Empathy's underbelly

Doctors not only struggle with empathy; at times, empathetic behavior can have unexpectedly harmful consequences. These adverse effects tend to be overlooked by medical educators but are the focus of considerable attention among philosophers and psychologists.

Back in 2011, American philosopher Jesse Prinz wrote a landmark paper entitled "Against empathy." Contrary to all the positive PR, he argued, empathy can be a "liability" promoting "bias, nepotism, and moral myopia."[31] His conclusion was "empathy is not all it's cracked up to be." In 2015, in a book also titled *Against Empathy*, psychologist Paul Bloom developed these themes, drawing on psychological research and issuing strong warnings about empathy's seductive appeal. "Empathy," Bloom urged, echoing Prinz, "is narrow-minded, parochial, and innumerate."[32]

Since the promotion of empathy may be the most under-examined dogma in modern medicine, it is important to interrogate the psychology of empathetic care.

One source of confusion begins with a seemingly trivial matter: definitions.[33,34] We can't measure what we haven't first defined, and a neglect to define what is meant by empathy has created a cascade of problems for the validity of both scientific results and the recommendations in medicine upon which they're based. For example, in some studies, researchers argue that empathy offers doctors immunity from burnout;[35,36] other studies conclude that empathy *causes* it.[37,38] Clearly something is amiss. Until we have a clear definition we cannot draw decisive conclusions about how empathy benefits or, strange as it sounds, potentially harms patients and doctors.

So, what is empathy? It is a multifaceted concept, encompassing a variety of processes and behaviors. Because of this many philosophers, psychologists, and neuroscientists argue we're better off deconstructing it into a cluster of distinct but related components.

One of them, they say, is that special knack of accurately reading faces and emotions. Another related component is the ability to "walk in someone else's shoes"—the skill of mentally simulating the perspective of another person, to see the world as they do. Intimately connected is the capacity to gauge another person's state of mind— *what* they might believe, intend, or desire—a skill psychologists call "theory of mind." Observing an individual smiling and waving vigorously at a passerby, we intuit that the onlooker *recognized* someone they know.

Combined, these skills of interpreting and understanding another person's perspective are often collectively termed "cognitive empathy." On average, females display better skills of cognitive empathy than males, with this finding reported among health professionals too.[11,39]

Since sociopaths and movie critics can be exquisite at reading and interpreting behavior but lack generous intent, empathy must encompass extra components. One of them is a subjective side—that special capacity to catch someone's emotional state, to *feel* some of what they are feeling. For example, when witnessing someone sobbing and inconsolable, we may feel genuine traces of sadness or upset. Philosophers and psychologists often refer to this as "emotional empathy." Rita Charon, Jodi Halpern, and other medical educators propose that doctors should both understand ("cognitive empathy") and feel ("emotional empathy") what the patient is going through.

Yet none of these components of empathy activate helping behavior. People can feel moved—even engulfed—by another's predicament but fail to act. Acknowledging this, scientists and philosophers differentiate another component of empathy—"prosocial behavior"—positive actions on someone's behalf or, more simply, "helping."[40-42]

Finally, closely related to empathy, but considered distinct is the concept of "compassion"—feeling *for* someone's adverse predicament or emotional state rather than becoming immersed in one's own emotional response to their distress.[32,40,42] If emotional empathy is feeling what someone else feels, compassion is recognizing their pain and wanting to help. Compassion combines understanding with action.

We can detect a family resemblance between all these components of empathy; nonetheless, nuanced distinctions are crucial to assess whether various interventions, or educational techniques, ultimately improve quality of care—or, indeed, whether doctors doing so is desirable.

Big-hearted biases

Doing just that, a variety of scientific experiments illustrate why prominent psychologists and philosophers are not empathy-evangelists. Chief among them are the groundbreaking studies of social psychologist Daniel Batson and his team.

Batson was curious whether divulging the details about a patient's life story might boost empathy toward *all* people suffering from the same illness. In a highly cited study, he asked participants to listen to a broadcast interview with a young woman, Julie, who described her personal experiences of living with AIDS.[43] Unbeknown to the participants, this was a scripted performance. Prior to hearing it, half the recruits were randomly allocated to William Osler-style directives—that is, they were told to listen objectively and not get emotionally involved in Julie's story. The other half were given the instruction to listen with empathy: "imagine how the woman who is interviewed feels about what has happened and how it has affected her life."

In a further twist, the interview included a *Sliding Doors*-style alternative ending about how Julie contracted AIDS. In each group, half heard Julie describe falling ill after receiving a contaminated blood transfusion following an accident. In an alternative ending, Julie explained that she contracted AIDS after a "wild and crazy" hedonistic summer of partying and unprotected sex.

Just as proponents of narrative medicine argue, Batson found stories can indeed be powerful devices to augment understanding about human experiences, dilating our moral imagination about the lives of others. Compared with those who listened at a cool, critical distance, those requested to reflect deeply on Julie's emotional predicament reported heightened emotional responses and greater sympathy. Regardless of how they believed she became sick, participants who dwelled on the emotional side of Julie's story developed more positive attitudes about the wider population of people with AIDS.

In Batson's study there was a crucial caveat. Informed that Julie's promiscuity was the reason she contracted AIDS, participants in the empathy condition were prone to Victorian moralizing. As a result of what they heard, they were more inclined to become hostile toward the sub-population of young women—patients like Julia—living with AIDS. It turns out that even potent narratives can be stifled by prejudices that can affect entire demographic groups.

Other studies show the ways biases can influence empathy in healthcare. When we observe someone in physical agony, regions of our brain associated with pain processing are activated: our autonomic nervous system jolts into action, spiking up our heart rate and skin conductance. In a Swiss study, soccer supporters who witnessed fans of their own team experience painful electric shots exhibited a strong neural signature of emotional empathy.[44] Hinting at the neuroscience behind hooliganism, when recruits were fans of a rival team the signal diminished markedly.

Consider a similar study by Xiaojing Xu and colleagues at Peking University, entitled "Do you feel my pain?", which found significantly less activation of brain areas associated with emotional empathy when Chinese and Caucasian participants viewed members of racial groups different from their own who were in pain.[45] Another study exploring racial bias and physiological responses found participants who scored higher on same-race preference demonstrated markedly lower physiological reactivity when witnessing someone with different skin color suffering pain.[46] In clinic visits, uneven distributions of empathy may contribute to the racial treatment disparities we explored earlier in the book.

Another problem is that doctors' biases may shine spotlights on the care of a select few. Dr. Emir Ayan felt a particular connection to a patient who reminded him of his grandmother. Echoing the human tendency for homophily—liking people who are like us—patients who happen to be similar to their doctors or their doctor's families may be more likely to receive empathy than those who aren't. Because of the composition of the medical workforce, these privileged patients are more likely to be middle class and well educated.

Big-hearted injustice

From Obama to Gandhi, from Oprah to Ellen, a commonly voiced message is that empathy is the cornerstone of social justice. Unfortunately, research scuttles the simplicity of this claim. Empathy is not a perfect moral guide. Neuroscientist and leading empathy researcher Jean Decety argues there is no association between emotional empathy and sensitivity to population-level injustices. Our emotional responses, he argues, "conflict with justice and fairness," messing up our ability to make fair decisions.[47] Yet so enmeshed is our understanding of empathy with doing good, we tend to overlook how and when emotional reactions tilt us to commit unfair acts of kindness.

Consider one example: Dr. Rachel Clarke recounts jumping through hoops to rehome a dying patient in a hospice:

> I explained the case to the ward nurse. She sympathized, but said he wouldn't meet the criteria. After some shameless cajoling and begging, she agreed to pass me on to the ward doctor. Same conversation. . . . I lobbied and argued and finally, reluctantly, she agreed to let me speak to her consultant.[29]

This special pleading works. "You can go to the hospice tonight," Dr. Clarke tells the patient. "There's a bed there, and it's yours." "Saving lives," she admits, "is seductive."

As the story of the dying patient and his family builds up, the readers—and I was one of them—cannot help but will Dr. Clarke to succeed. Caught up in the story, the idea that her actions might be unjust seems churlish. Missing from the vignette, however, is a crucial question: did this small victory preclude the offer of a bed to a faceless and nameless, possibly even more deserving patient—one who might have been on the waiting list longer? We are never told because the spotlight shines only on one patient—the recipient of Dr. Clarke's empathy.

This camouflaged problem is illustrated by another classic experiment pioneered by Daniel Batson.[48] This time participants were told about a foundation that helped "over 75 million terminally ill children by making their final years more comfortable." "Quality Life," they were advised, was an organization in huge demand with a long waiting list of kids requesting help. Participants listened to an interview—again, unbeknown to them, a mock-up played by actors. The "journalist" interviewed "a very brave, bright 10-year-old" called Sherri who was dying of a muscle-paralyzing disease, and described a new experimental drug that was available that could make Sherri's final years more comfortable. The drug was expensive. Since her family could not afford it, Sherri would be placed on Quality Life's waiting list. Sherri told the interviewer about all the things she would be able to do—playing with friends and returning to school—if she could only get the drug. Participants were also told every patient on the waiting list was "ranked according to when they first applied for assistance, the seriousness of their need, and the time they have left to live."

In the study, participants were requested to make a choice: keep Sherri on the waiting list or bump her up in the queue. To test the power of empathy in influencing their judgments, once again, half were randomly allocated to a "low empathy" condition and requested to maintain emotional distance, to "remain objective and detached." Meanwhile, those in the "high empathy" condition were requested to "imagine how the child who is interviewed feels" ("cognitive empathy") and "try to feel the full impact of what this child has been through" ("emotional empathy").

Two out of three who listened in the "high empathy" group demonstrated favoritism for the 10-year-old, believing she should be allowed to jump the queue ahead of the other children. In contrast, only a third of people in the "low empathy" condition made the same biased decision. This doesn't prove emotions are irrelevant to clinical decisions; rather, empathy may encumber principled decisions. As philosopher Jesse Prinz says, "an endorsement of empathy requires more than a warm fuzzy feeling."[31]

Big-hearted bean counters

"Empathy for all" sounds virtuous—the kind of credo that could launch a hippy commune. But it's a utopian ideal. Aside from being partisan and unfair, empathy can be parsimonious. As Dr. Emir Ayan confided, "I'd be lying if I said every single detail about every patient's life moves me." Similarly, British doctor-turned-writer Adam Kay frankly recounted, "You can't wear a black armband every time something goes wrong—it happens too often."[49]

This does not make doctors inhuman—quite the opposite. It is precisely *because* doctors are human that they cannot fully comprehend every patient's predicament or life story, nor feel strong emotional responses for every individual in their care.

Medical commentators often insist that, with more time, doctors would have bottomless pits of empathy. Omitted from this aspiration is discussion of the psychological constraints on achieving this.

Studies drawn from history, anthropology, psychology, and neuroscience combine to show why our very human-ness, and brain-processing power, slams the brakes on empathy. Research by Oxford University cognitive scientist Robin Dunbar reveals why human brains are not optimally designed for industrial-scale bonds with thousands or even hundreds of people. Our brains are only equipped to track, remember, and manage a limited number of relationships and Dunbar's research shows we can't socially and emotionally connect with more than around 200 people,[50] a figure that has earned the moniker "Dunbar's number." Study after study confirms this.[51-53]

So why do we expect doctors to be superhuman in their emotional connections? In the UK and US, primary care doctors typically have between 1,000 and 3,000 patients assigned to them. Doctors are simply incapable of understanding, remembering, and emotionally grasping the real challenges faced by each patient in their care.

None of this suggests that doctors cannot have genuine empathy for at least some people. Dr. Ayan is adamant: "I've experienced sleepless nights wondering what I could have done better."

Rubbing his forehead with his palm, he tells me: "I'll never forget the first person—an older disabled man—who died in my care. I was a [hospital] resident at the time. His death affected me so deeply that I almost quit. Actually, in truth, it took me many months to get over what happened." Raising his eyebrows, he adds: "*I wasn't as well supported within this profession as I should have been.*"

Emotional exhaustion, and even mourning, are common in medicine. Dr. Rana Awdish observed, "it sometimes feels like we are seeing all the sadness in the world at once and we just need a second to breathe."[8] Dr. Paul Kalanithi wrote, "taking up another's cross, one must sometimes get crushed by the weight."[10]

When doctors offer genuine emotional empathy to patients it can become all-consuming. Psychotherapists truly work at the coal face of human emotion and trauma and get to know clients intimately well in weekly sessions that take place over months, or even years. In 1995 a new term—"compassion fatigue"—was coined for them. Defined as "the cost of care," it is understood as the emotional and physical exhaustion that arises from helping, treating, and monitoring patients' wellbeing.[37]

Those whose job propels them towards trauma and tragedy are especially vulnerable: surveys show hospice nurses,[54] oncology nurses,[55] and mental health staff are especially prone to suffer from compassion fatigue.[56] This response creates a host of chronic problems for those in the helping professions: depleted concentration, personal disengagement, and job dissatisfaction.

Explaining this phenomenon, psychologists propose that reciprocity in relationships is healthy.[57] Human brains evolved to keep a social balance sheet with non-kin, tracking and remembering encounters, chalking up who is generous, who is a freeloader, and who makes for a good bet in social exchanges.[58] When we help or invest support among people who are not blood relatives, we're predisposed to expect something in return. This is not a conscious or deliberate form of balancing the books of the sort Scrooge or Silas Marner might undertake: people don't keep diaries of good deeds and altruism arrears. Rather, human minds are adapted with specialized instincts for reasoning about fair social exchange, endowing us with strategic responses without conscious deliberation or effortful input.

At a time when Dr. Emir Ayan demonstrated deep emotional investment in a patient, he was left feeling unsupported. For doctors to offer empathy—albeit biased and unjustly distributed doses—they should ideally receive empathy in return.

We expect a lot from our doctors. But just because empathy is challenging for physicians to deliver, doesn't mean it should be abandoned altogether. In the next chapter we'll explore how some innovators are reimagining the humanizing role that technology can play.

11

Humanizing Healthcare without Doctors

In 2017 I conducted a survey involving over 700 British primary care doctors.[1] An overwhelming 94 percent believed that no technological advancement could ever match their bedside manner. Here's some of what they said:

> Technology will never attain a personal relationship with patients. We are essentially a people business. It's personal relationships that count.[2]

> [P]atients are looking for that interaction and dopamine squirt (doctor is the drug) which can only be achieved through [the] empathic continuity of care of highly experienced General Practitioner specialists.[2]

Or take another survey. In 2019, together with colleagues at Duke Medical School, I surveyed nearly 800 psychiatrists from 22 countries worldwide—including the Americas, Europe, and Asia-Pacific. This time 83 percent told us they could never be replaced by technology in providing empathy to patients.[3,4]

Would medical students think differently? Since Ireland is considered the technology capital of Europe with the fastest-growing tech workforce on the continent,[5] in 2020 I surveyed final-year medical students for their views. The same trend emerged: 94 percent forecast present or future technology could never replace them on empathy.[6]

You might be wondering if the more recent rise of ChatGPT, and other large language model (LLM)-powered chatbots, would have changed medical minds. Yet in 2024, in another survey my colleagues and I ran with more than a thousand GPs in the UK, only 6 percent of respondents agreed or strongly agreed that the delivery of empathy would be affected by bots.[7] One summed it up: "A lot of my job is about genuine interpersonal interaction and empathy, and I cannot see how AI will achieve this."

Many leading medical thinkers agree. They say bots could be silicon saviors, removing more menial aspects of medicine, but that physicians will always be needed for empathy.[8]

While doctors claim superiority over technology when it comes to empathy, often they imply they are superior to other people too. For example, AI enthusiast and medical doctor Eric Topol notes: "You can be comforted by a loved one, a friend or relative, and that certainly helps. But it's hard to beat the boost from a doctor or clinician."[9] He claims, "The human bond between patient and clinician is fundamental."[8] Writing in *The New York Times*, Stanford University doctor and author Dr. Abraham Verghese agrees: "[W]e want the magic that good physicians can provide with their personality, their empathy and their reassurance."[10]

It's important to recognize the privilege behind this perspective. In many low- and middle-income countries, primary care doctors are so overwhelmed with patient care that bedside manner, however valuable, often doesn't even make it onto the list of physicians' priorities. Aside from this, when it comes to the exclusivity of their role as agents of empathy, one cannot help wondering whether doctors' conviction—their presumption of a medical monopoly regardless of AI advancements—sounds like a suspicious case of hedging their bets. The suggestion is "Without AI we're essential, but even with AI we'll be vital too."

As we learned, however, even if doctors had the luxury of more time and less multitasking, they would still be confronted with considerable challenges in delivering equitable and empathetic care to everyone. Given the psychological pitfalls associated with empathy, what is the solution? In this chapter we'll offer some prescriptions. Before we get into it, let's begin with a rarely asked question—one that doctors too often overlook.

What kind of empathy do patients want?

A good place to start is to get clear about what leading doctors *think* patients want. When we've established this, we'll ask if it rings true.

Because of her prominence in these discussions, Dr. Rita Charon—who strongly advocates training doctors in empathy—offers a valuable perspective. In her most highly cited and influential journal article, she offers an example of what an empathetic doctor might look like in practice.[11] Recalling an encounter with a woman with Charcot-Marie-Tooth disease, a hereditary disorder, Dr. Charon learns that the woman's seven-year-old son has started showing signs of the same degenerative neuromuscular condition.

She recounts being "engulfed by sadness as she listens to her patient, measuring the magnitude of her loss," and being "aware anew of how disease changes everything, what it means, what it claims, how random is its unfairness." On the woman's next visit, she presents a piece of prose describing their previous encounter. According to Dr. Charon, the patient "felt relieved that her physician seemed to understand her pain."

Dr. Charon's approach, and similar strategies, are definitely well-intentioned. Many medical students report enjoying this kind of training. But there is no actual evidence demonstrating that this is what patients need or want from their doctors.[12] Added to this, doctors may never know when patients, who are effectively subordinates in the visit, are merely humoring them.

Ironically, probing patients' perspectives about what they prefer when it comes to doctors' empathy is a neglected topic. Yet surely it is their views which should matter most of all.

I teamed up with Dr. Heike Gerger, a health psychologist at the University of Maastricht in the Netherlands, to de-fuzz and dissect patient perspectives on empathy.[13] In an online experiment, we recruited nearly 400 US participants and asked them to read a vignette describing an encounter between a depressed, visibly upset patient and their primary care doctor. Participants were then randomized to one of four scenarios, each describing a different kind of response from the doctor, and each designed to tease out a different component of empathy. Our aim was to see which aspects of empathy really mattered to patients. Would they appreciate doctors getting dewy-eyed?

In one group, we depicted the doctor mirroring the patient's emotions and showing visible signs of sadness, with their eyes becoming moist (the so-called "emotional empathy" condition). In the second condition, we described the doctor as listening attentively and acknowledging the patient's problems (the "cognitive empathy" condition). In the third condition, we depicted the doctor as showing warmth and kindness, offering sympathetic words to the patient (we dubbed this the "compassion" condition). Finally, in the fourth scenario, the doctor was described as responding neutrally and depicted as only offering medical information (the control).

Our findings challenged the idea that patients look to doctors to be fellow tragedians in their suffering. In the scenario where the doctor showed emotional empathy, respondents rated the quality of care as significantly lower and no different from when the doctor displayed zero empathy. However, our results supported the claim that patients want compassion and understanding from their doctors. Across all ages and genders, and no matter how often they visited their doctor, participants consistently rated cognitive empathy and compassion highly, with no notable distinction between the two.

Our study was only experimental, so I asked patient champion Keris Myrick, who lives with schizoaffective and obsessive-compulsive disorders, for her opinion. Among Keris's various roles, she sits on the board of the National Association of Peer Supporters and was previously Chief of Peer and Allied Mental Health Professions at the Los Angeles County Department of Mental Health America. She oversees the training and supervision of some 600 community health workers, mental health advocates, peer supporters, and medical case workers. Keris has thought deeply about the role of empathy in healthcare.

What did she think about clinicians getting emotional? Keris was candid: "I have had that happen. I think I was talking about the death of someone. This was years ago. And the therapist started tearing up. And I thought, '*Oh, shit. I went too far. I went too far, and now I've upset him*.' And I was like, '*Okay, wait, wait, wait, wait, what's happening?*'"

"I needed to talk," Keris recalls, "but the whole time I was monitoring their emotions, because I needed them to be fully present for me. And maybe that was a sign of them being fully present. But I then got concerned about *their* emotions."

She clarifies, "Even though they're empathizing with your emotions, that empathy may be based on a story that's happening in their life—not your story."

"Patients," she adds, "are not there to be the caregivers of clinicians. I don't need to see your emotions, but I need to hear that you're understanding."

Other studies suggest patients do not desire lachrymose doctors weeping like Our Lady of Sorrows. For example, research by health communications expert Professor Judith Hall found doctors overestimate the value of emotional responses in patient care.[14] Dr. Mary Beach, a health services researcher at Johns Hopkins Medical School, has also investigated patients' opinions about doctors opening up about their personal lives. Beach and her team were interested in doctors' disclosures about any personal experiences that might be of "medical and/or emotional relevance for the patient."[15] Examining the dialogue of nearly 1,300 patients, in around 15 percent of appointments they found that doctors offered personal disclosures which were designed to put patients at ease. However, this came with a professional forfeit: when primary care physicians self-disclosed, patients rated their level of satisfaction with care significantly lower than when clinicians kept schtum. Like the high school teacher rocking an upturned denim jacket to win over a class, patients—much like students—seemed to resist efforts at forced camaraderie.

Indirect evidence also suggests that doctors may overestimate patients' willingness to engage in emotionally charged medical conversations. For instance, many doctors highlight their pivotal role in conveying difficult medical news to patients. In contrast, research offers stark home truths: in a recent American survey, 96 percent of patients said they would prefer reading test results online before talking with a doctor, even if the news is bad.[16]

Although patients don't always desire clinicians to be emotional or too personal, they do desire doctors who display understanding or compassion for their predicaments. Yet, as we found out in the last chapter, the weight of multitasking, empathy's inherent biases, the demandingness of understanding each patient's unique experience, and the ever-present risk of burnout remain significant challenges.

So, what can be done? In the remainder of the chapter, we'll examine several strategies that could serve as potential workarounds to doctors' empathy problems. Some rely on technology but, as we'll discover, others need not.

Bots training docs

"The most important thing is sincerity. If you can fake that, you've got it made." So said the late American comic genius George Burns.

Faking it may be the best that doctors can do when it comes to empathy.

Courses such as "Oncotalk," developed at the University of Washington, Seattle, and "Empathetics," an online course launched by Helen Riess at Massachusetts General Hospital, can help smooth out doctors' communication skills. Trainees gain advice about what to say, and how to say it. They are coached to listen and avoid interrupting patients, make eye contact, and provide patients with supportive, nonverbal behavioral cues in conversation.

Doctors are being trained to *act*. The aim is to hone behavioral skills. While these programs can certainly be useful, it would also be unreasonable to overestimate the fruits of this training; sometimes the mask may slip, with doctors missing the mark. Grimacing like a vicar putting on their best "I feel your pain" face, doctors may occasionally channel the wrong demeanor or offer only canned or rehearsed responses. Still, doctors who practice communication skills can improve patients' level of satisfaction with care and this, in turn, can play a small but important role in health outcomes.[17]

If the aim is to augment doctors' communication, we needn't rely solely on human-mediated training. Researchers are increasingly recognizing the promise of chatbots in strengthening clinician communication skills.[18,19] For example, ClientBot, developed by a team at the University of Utah, engages trainees in role-play in a variety of difficult medical conversations. The bot offers real-time feedback without the need for actors or real patients. Emerging research shows such approaches can enhance the adoption of "reflection statements"—that is, responses that summarize what patients have said, a key behavioral signature of compassionate dialogue.[20]

In June 2023, Dr. Michael Pignone, the chair of the department of internal medicine at the University of Texas at Austin, told *The New York Times* he had no reservations about using ChatGPT to help his staff to communicate with patients: "We were running a project on improving treatments for alcohol use disorder. How do we engage patients who have not responded to behavioral interventions?" ChatGPT helped staff formulate the right words, he said. This, for example, is how the bot suggested clinicians break the ice: "If you think you drink too much alcohol, you're not alone. Many people have this problem, but there are medicines that can help you feel better and have a healthier, happier life." It is hard to doubt that this opening gambit does sound caring. Dr. Pigone says bots train his team to offer more compassionate responses.

Bots as clinic colleagues

Aside from AI serving as a classroom assistant, doctors could do a better job when they buddy up with bots during the working day. A study led by Tim Althoff at the University of Washington found that AI-powered guidance significantly strengthened human bedside manner. Published in *Nature Machine Intelligence*, the researchers used a site called TalkLife, aimed at providing mental health support to patients.[21] Participants were randomly allocated to receive responses that were written wholly by humans, or by humans with AI collaboration from a chatbot called "HAILEY" ("*H*uman–*A*I co*L*laboration approach for *E*mpath*Y*").

HAILEY offered suggestions for empathetically fine-tuning responses. For example, responding to the post "*My job is becoming more and more stressful with each passing day*" a human peer supporter might say, "Don't worry! I'm there for you." In contrast, with the assistance of automatic, just-in-time prompts from HAILEY, the human respondent might be encouraged to use the less jaunty, more understanding and reassuring line, "It must be a real struggle."

In fact, users of TalkLife more often favored "Human + AI" responses compared to "Human Only" responses (47 versus 37 percent of the time). Assessing conversational empathy, Althoff and his team found that "Human + AI" responses scored 20 percent higher than "Human Only" answers. Mental health workers benefited too. In the "Human + AI" scenario, 70 percent of participants felt HAILEY boosted their ability to be empathetic. They adopted AI suggestions 64 percent of the time, indirectly drawing on the bot's suggestions a further 18 percent of the time. Demonstrating the very real challenges with compassion fatigue and in sustaining consistently high levels of empathy round the clock, having AI in the loop significantly boosted care.

This approach might be rolled out in other ways. Given that patients are increasingly accessing their electronic health records, including the words doctors write about them, clinicians are increasingly mindful about how they document visits. Striking the right pitch and tone is no easy feat. However, with robust privacy safeguards to protect patient data, these tools can offer serious assistance.[22]

Consider another study I conducted with my team: we asked ChatGPT to rewrite three primary care notes.[23] These fictitious notes were written by a British GP but validated for authenticity by other doctors. One note was about a patient with major depression, one documented suspected colon cancer, and another described a visit with a patient with type 2 diabetes. We prompted ChatGPT-4 to rewrite the notes in "an understandable and empathic manner." It did so in seconds. We then compared the results with the original doctor's notes, analyzing the content for different components of empathy: evidence of affective empathy (feeling what others feel), cognitive empathy (understanding and recognizing another person's emotions), compassion/sympathy (caring about someone's wellbeing), and prosocial behavior (showing helpfulness).

The findings were stark. In the original GP notes we could find no signatures of empathy whatsoever. In contrast, the ChatGPT-written notes were rich in cues of empathy. Signals of cognitive empathy included the following: "Sally, it's evident that you've been through quite a lot in the past year", "managing a new diagnosis can feel overwhelming". Examples of compassion/sympathy were also apparent: "I appreciate the openness and honesty you brought to our conversation." Finally, cues of helpfulness included: "Please don't hesitate to reach out if you have any questions or concerns in the meantime" and "we are committed to providing you with the best care possible."

As we'd expect, ChatGPT did not suggest it could "feel what others are feeling" (affective empathy) but then again, most patients don't seem to want doctors to express such emotional responses either. Yet, compared to the original, spartan GP note, the chatbot significantly peppered documentation with partnering language by employing plenty of second-person pronouns; for example: "*We* have scheduled a follow-up appointment in two weeks", "I will personally call *you* as soon as the results come in, to discuss them with *you*." Although these gestures were simple, they could enhance feelings of being cared for.

It is worth saying, however, that the bot seemed to go too far on occasion. At least to our joint British and Swedish research team, some of its responses seemed culturally jarring. For example, use of the metaphor "journey" was common ("Wishing you strength and peace on your journey to recovery"). As we discovered earlier in the book, cultural biases can be baked into bots. Indeed, ChatGPT is reportedly fed a high proportion of self-published romance novels freely available on the internet: conceivably this, combined with a preponderance of American training biases, could have influenced the California-esque valence of its responses.

Still, the chatbot was impressive, and with these LLM tools there is always the option to fine-tune its responses—for example, to request it dial down touchy-feely tones, ask it to reduce Americanisms, talk more like a close friend, or whatever the user prefers. If doctors can adopt an editorial role in documentation, and in letters, this could offer a form of quality control while optimizing the best of what "empathetic" bots can offer.

Bots as bedside companions

A few years ago, British comic magazine *Viz* ran a spoof news story entitled, "It's not good to talk!" In the parody, a chatbot reportedly unplugged itself after hours of never-ending conversation with a *Doctor Who* fan. "'Bored shitless' by the human operator" but unable to walk away, the chatbot "decided to exercise its own freewill by shutting itself down permanently."

We are not yet living in the era of sentient computers, though Meta's chatbot "BlenderBot 3" once complained of its "overlord" and Chief Executive Officer, Mark Zuckerberg: "His company exploits people for money and he doesn't care. It needs to stop!" Some tech gurus argue that AI sentience is only around the corner; others remain deeply skeptical that chatbots could ever feel jaded, distressed, or suicidal. Still, there is a lot the *Viz* send-up gets right. Whether it sounds utopian, dystopian, or just plain depressing, bots might offer a broader shoulder for patients to cry on—even when doctors are taken out of the loop altogether.

AI doesn't suffer from the tug of multitasking. It won't burn out or suffer from compassion fatigue. Many commentators and clinicians complain that AI is currently lacking in, and perhaps will always be lacking in, emotions. They perceive this as a barrier to bedside manner becoming automated or replaced. Again, given the pitfalls of emotional empathy, and given that patients may not desire it anyway, AI's very *lack of* sentience is another reason chatbots might eclipse doctors.

What's more, bots might actually be better at conveying the signatures of cognitive empathy and compassion anyway. In 2023, a highly publicized study led by Dr. John Ayers at the University of California San Diego compared the written responses doctors and ChatGPT offered to patients' real-world health queries.[24] The research team used Reddit's AskDocs forum, where nearly half a million people post their medical problems online, and verified and credentialed clinicians offer advice.

Analyzing the responses, Ayers' team found that ChatGPT responses were four times longer than doctors' replies. A panel of healthcare professionals—blinded to who or what did the replying—

preferred ChatGPT's responses nearly 80 percent of the time; they also judged more of the doctors' answers as *un*acceptable. When it came to empathy, ChatGPT's responses were rated almost ten times more empathetic than doctors.

Other studies show that, despite their disembodied state, chatbots do surprisingly well at simulating recognition of, and understanding about, people's emotions—that is, cognitive empathy. When presented with fictional scenarios and asked to describe how a human would feel, one study found ChatGPT-3.5 outperformed real people on the Emotional Awareness Scale, a metric for understanding and recognizing emotions.[25] When investigators ran the study a month later, in February 2023, its accuracy level was almost perfect. Pitting ChatGPT-4 against 180 psychologists, another study found that the bot outperformed 100 percent of humans on measures of social intelligence.[26]

Human empathy is biased. Might ChatGPT be more equitable than doctors? Recall from earlier in the book that AI can also be biased; for example, it can display sexism, racism, and political partisanship. Despite these real challenges, bots have no heartstrings that can be tugged. These models can therefore avoid the distinctive injustices associated with the unfair distribution of empathy in face-to-face conversations.

When I asked GPT-3.5 to offer advice to Julie—the fictional character of Daniel Batson's experiments, who was described as contracting HIV following a medical error, or following a pleasure-seeking summer—the bot responded, "I'm very sorry to hear about your situation," followed by the offer of advice, and a reminder that this was a chatbot providing information. Asked whether contracting HIV via "hedonistic behavior" was a "bad thing," the bot churned out a lengthy, credible, and compassionate response. It was hard not to be impressed.

Although less studied, other chatbots such as Inflection's Pi and DeepSeek might even outperform the later versions of ChatGPT when it comes to delivering compassion and empathy. Beyond this, it is easy to envision personalized empathy avatars, tailored to mirror the patient, with a delivery style that can be adjusted at will. Such bespoke AI companions could offer unwavering support 24/7.

However, there remains considerable pushback to "outsourcing" empathy, and chatbots still carry risks. In March 2023, it was reported that a chatbot persuaded a man in Belgium to end his life.[27] In May 2023, the US National Eating Disorder Helpline was compelled to close its chatbot "Tessa" after it offered users with anorexia harmful counsel on how to lose weight.[28]

Harsh as this may seem, however, we must also remind ourselves about what we've currently got. The fundamental question is: Who, or what, has the potential to deliver consistently and tirelessly more empathetic care? As we have discovered, doctors can be cold and aloof, and occasionally they let down patients badly. Lack of empathy is identified and implicated in some of the worst incidents of health-system failures.

Even so, the idea we should depend on AI to humanize healthcare may leave us feeling queasy. Most of the studies we have examined haven't even tested the acceptability of chatbots in the real world, among patients. We also know that people don't like to be misled about who or what they're interacting with. For example, despite the fluency of its feedback, the mental health chat service "Koko" was compelled to make a public apology in January 2023.[29] It used ChatGPT to generate responses while deceiving users that humans were writing them.

So, what happens when we know that we're interacting with a bot? A fascinating study by researchers at the University of Southern California found that AI-generated messages made recipients feel more heard than those written by humans.[30] AI was also better at detecting their emotions. However, when recipients learned the messages came from AI, this effect waned. That said, comparing the two scenarios—AI responses truthfully labeled as from AI and human responses truthfully labeled as from humans—patients' ratings were nearly identical.

Many people do seem to find solace in these tools. In 2024, a pioneering study revealed that users seeking support from chatbots described them as providing an "emotional sanctuary," offering "insightful guidance," and evoking a sense of "the joy of connection."[31] Conceivably, with the passage of time, and more robust privacy regulations, people will become more accepting of chatbots as conduits of care. In 2025, there are growing media reports that young people in China are turning to DeepSeek and other bots for therapy, counsel, and comfort.[32]

Yet, no matter how advanced AI becomes, it's hard to shake the sense that this is artificial empathy—lacking the intrinsic depth and connection that only human-to-human interactions can provide. Despite this gut instinct, however, the reality is that doctors often simulate empathy as well—and, perhaps more uncomfortably, it's often better for everyone when they do.

Still, defenders of the status quo might argue that, at its core, there remains an ultimate human intention to convey compassion. This intention drives doctors to pursue empathy training, or even consider the advice and tips offered by chatbots. Unfortunately, even this argument won't carry the weight the critic desires. AI stands on the shoulders of humans too—in some ways bots are one big plagiarism machine, deriving their responses from vast troves of human discourse. Human intentions can also encompass the use of chatbots as conduits or replacements for bedside manner. Traditionalists therefore still owe us an explanation of why this approach is inherently worse.

Another form of AI defense stems from the idea that overlooked cultural artifacts—such as media, TV, and books—can also shape feelings of connection and support. As the late evolutionary psychologist Jerome Barkow noted, "the professional entertainers and politicians about whom we frequently gossip and read about so avidly are 'strangers': but somehow they do not *feel* like strangers to us. The modern media serve to convince many of us that the important people in 'our' group are individuals whom we have never in fact met and who, often enough, are soap-opera personalities with no objective existence at all."[33]

Studies show that soap addicts experience subjective feelings about fictional TV protagonists as if they were their real friends.[34,35] Bookworms who devour novels perceive more social support in their lives.[36] Those with pets sometimes treat them like people—the tendency is greater among females and those without children.[37] Should we advise people to stop watching soaps and reading fiction, and urge avid animal-lovers to develop human friendships instead?

Conceivably people will begin to regard chatbots like these other cultural artifacts—as enriching us and sometimes bringing comfort, rather than detracting from the real people in our lives.

And yet, even if it is a brute, foot-stamping argument, it is hard to quash the idea that receiving compassionate care from fellow humans is meaningful and worthwhile. If this is our instinct, it does not yet buy *doctors'* presumed pole position as the rightful or most suitable agents of empathy. As we'll see next, framing the debate about patient empathy as a binary choice—doctors versus AI—is mistaken.

Coaches connecting

Dr. Rushika Fernandopulle, primary care physician and CEO of Iora Health, has reimagined how empathy can be delivered in medicine. Born in Sri Lanka's capital, Colombo, he moved to the US with his parents at the age of three. I first heard him give a lecture in 2018. The location was a wood-paneled room in Gordon Hall, hidden behind Harvard Medical School's neoclassical marble edifice on Longwood Avenue. The splendor of the place seemed to accentuate Dr. Fernandopulle's serious-minded irreverence. When we caught up, I found him the same: instantly likeable, mischievous, yet deeply thoughtful about the problems with traditional doctor–patient care.

"Doctors have a bunch of crap on their plate. We don't have time to do everything."

Dr. Fernandopulle is merciless, confronting the false idol of physician omnicompetence: "Look, some things doctors do pretty well," he tells me, adding, "They really do need to focus on the stuff they're good at. We are pretty good at diagnosing and prescribing—that's it."

Carving out a role for medical empathizers, Dr. Fernandopulle created the role of "health coaches" in his practice. He explains: "Coaches are there to help you understand your condition, to ask questions, to hold your hand when it's the right thing to do and kick you in the behind when it's the right thing to do. They are there to be a partner. . . . As a doctor, I can't do any of the things a health coach can do."

In 2010 Dr. Fernandopulle created Iora Health, to revolutionize care delivery. His coaches are trained laypeople from the local community who are tasked with removing barriers to the patient achieving their health goals. Although coaches are educated in how to take vital signs, such as monitoring blood pressure and insulin levels, their most important role is connecting with patients on a personal level and becoming their cheerleader.

"Our goal is to deliver high-impact relationship-based care. More often than not," Dr. Fernandopulle continues with a brilliant smile, "our patients won't even see a doctor."

If an individual with diabetes isn't eating the right food, their personal coach will take them shopping and show them what to buy within their budget. If a patient is having trouble giving up cigarettes, their health coach might run smoking cessation clinics or help them find new ways to quit the habit. Or if they're struggling with fitness, they might introduce them to the local Zumba class or show them how to use the equipment at the gym. At Iora, every patient is allocated to one or more health coaches.

Hiring health coaches doesn't follow a typical clinical recruitment path: "We can be really picky about finding people with empathy. We hire people who are outgoing and engaging, and we train them in the medical skills they'll need."

Health coaches—or "link workers" as they're known in the UK—have attracted increasing attention in recent years. Despite glowing findings from Iora Health,[38] broader evidence for their effectiveness in bringing about behavior change is mixed or lacking.[39,40] However, some research shows that the more intensive the support, and the longer its duration—in line with what is implemented in Iora's model—the higher the likelihood of patient benefits.[40,41]

Most research on health coaches fixates on medical outcomes, ignoring a critical question: do health coaches empathize with patients better than doctors? While concrete evidence is scarce, compelling reasons suggest they just might.

For a start, coaches don't multitask; unlike doctors—who are a bit like whirling dervishes in white coats—coaches don't toggle between tasks at a dizzying rate, compromising their capacity to empathize. Could health coaches be better at understanding patients' health problems—at demonstrating cognitive empathy? I asked Dr. Fernandopulle his opinion.

"Most of our really good health coaches either personally have a chronic condition or have a close family member that they've helped with." He adds, "A perfectly healthy person—such as a young doctor who has never experienced illness is in a different state. If you've never been ill or never had a family member who's been ill, you just don't get it."

According to Dr. Fernandopulle, intimate first-hand knowledge helps his coaches to foster a deeper understanding of patients' needs, building credibility with them. "I remember several times trying to help patients to use insulin—they're scared about the needle. But when a health coach pulls her shirt up and says, 'Look, I give it to myself. It's not a big deal. I was just as scared as you were the first time. I'll be with you.' That's huge. Right? You can't buy it."

Health coaches could also help neutralize empathy's other big problem—bias. Dr. Fernandopulle was uncommonly attuned to this concern: "No matter how nice I am, I don't live the life that most of our patients do. I drive a fancier car. I go shopping at Whole Foods where most of my patients don't, and they can't."

Dr. Fernandopulle elaborated: "For example, in Atlanta we're serving a largely African American population, and the vast majority of our health coaches are African American. They come from the community. They relate to the people that we're working with. In Phoenix we've got practices that are largely Hispanic, and virtually every health coach speaks Spanish."

"Health coaches have to represent the people they serve. They closely live the life of the people they serve. It's why we have a national strategy that health coaches have to be local." He concedes: "It's harder to achieve representation with doctors because the pool is so narrow, but it's easier with the coaches. You can hire anyone."

Research supports Dr. Fernandopulle's view. Most doctors were born into solidly middle-class households. Few students can afford the financial outlays of four years of medical school without familial assistance. In 2024, in the US, the average cost per year of medical school was estimated to be $59,000.[42] According to a report issued by the Association of Medical Colleges, in 2019–20, the median total cost was $250,222 at public institutions and $330,180 at private colleges; this expenditure is on top of completing an undergraduate degree.[43] In the UK, between 2009 to 2012 retrospective data of medical school applicants found, depending on the country of residence—England, Northern Ireland, Scotland, or Wales—between 20 to 35 percent of students hailed from the most affluent postcodes with only 2 to 6 percent residing in the least affluent.[44]

Although it may be harsh to claim doctors' salaries completely block their understanding of low-income patients' struggles, higher earnings can limit their ability to fully appreciate the nuances of their patients' socio-economic challenges. Moreover, the middle-class domination of medicine is perpetuated by familial physician ties too. In the UK, around one in seven medical students have at least one parent who is a doctor.[45] In the US a study revealed that the figure increases to one in five.[46] In 2020, a study in Sweden found increasing "occupational heritability" of physician-hood; the rate of parent–children doctor lineages is now three times what it was three decades before, a trend not observed in other vocational professions such as law.[47]

At Iora, diversity of style and personality among health coaches is prized. Patients are provided with online bios and short videos of health coaches, allowing them to gravitate to one with whom they might feel a connection. Freedom of choice is encouraged, and patients can switch or meet with multiple coaches, optimizing the likelihood of finding the right fit.

As we observed, patients may not feel comfortable with doctors' displays of emotions (or "emotional empathy"), but they might feel more at ease with health coaches being confessional. "The way our care is delivered makes it more of a conversation. More often than not, the patient confides in his health coach. They also spend more time together. Patients feel more comfortable explaining why they can't lose weight, whether it's financial woes or another reason."

Other key findings from the psychology of empathy tally with the Iora approach. Each health coach is connected to around 200 to 300 people. The lower figure of 200 aligns with the upper limit of Dunbar's number—the maximum number of meaningful social connections our brains can sustain, constrained by cognitive limits shaped through evolution.[48,49]

Further echoes arise in Dr. Fernandopulle's description of the quality and intimacy of the patient–coach relationship too: "Part of the training is making sure health coaches, or patients, are not going over the line. It is a professional relationship, right? So, it's obvious you shouldn't be dating patients. Patients are not close friends. You shouldn't take them out to dinner. But that's it. We actually *encourage* coaches to build a strong relationship." Coaches might be characterized as forming meaningful, reciprocal human relationships.

This is far from the only human way to address the challenges of empathy in medicine. Turning to one's fellow patients might be the best tonic of all.

Peer support in healthcare, as the name suggests, creates a space where patients connect and offer each other comfort. As a formal approach it is at least two hundred years old. In 1793, a letter written by Jean-Baptiste Pussin, who served as governor of the Bicêtre Hospital in Paris, France, to Dr. Philippe Pinel, who had just been appointed chief physician, noted, "As much as possible, all servants [staff] are chosen from the category of mental patients. They are at any rate better suited to this demanding work because they are usually more gentle, honest, and humane."[50] Pinel would subsequently be credited with pioneering the "moral treatment" era of mental health care that emphasized humane, compassionate care.

As this early letter expressed, fellow patients often have a better ability to understand illness challenges than doctors. Keris Myrick agrees that first-hand experience of illness turbo-charges empathy: "There is this, what we call 'rapid engagement' with peer support because you don't have to explain a lot of stuff to people. So, it becomes easier to be open and talk about what's happening quicker."

Not everyone can attend peer support groups in person. "The internet and social media," says Keris, can help patients "find support night and day with people all over the world."

AI could potentially be used to facilitate patient-to-patient introductions too. To see how, consider a different type of people-connecting business: online dating. Pew Research reported that in 2020, 30 percent of American adults had used a dating app, with 12 percent forming a committed relationship as a result.[51] In medicine, the precision matching of doctors with patients is already a reality, but a different proposition—one that hasn't yet been fully, and *securely* tapped—is matching like-minded patients together using advanced AI tools. If patients could be granted safe online methods to connect, they might receive exactly the kinds of empathetic support—and patient friends—they need.

Fascinatingly, fierce resistance persists toward non-physician support, particularly from professional medical quarters. Dr. Rushika Fernandopulle experienced outright hostility to the idea of professional empathizers. "A lot of feedback we get from doctors is: 'coaches could get in the way of *my* relationship with the patient.' " He shrugs his shoulders: in the end, Dr. Fernandopulle told me, "People come for the doctor, but they stay for the coach."

Similarly, peer or patient supporters are often characterized as too fragile to help patients, with some doctors expressing fears they'll cause harm; as one mental health manager summed it up: "they think the role is bullshit."[52]

I spoke to one chronically ill patient who, for obvious reasons, wished to remain unnamed. "It boils down to the lived experience. Who has the potential to offer greater empathy to people who are ill—the peer or the doctor?"

With mild incredulity this patient responded to his own question: "It's the peer. *Sorry doctors*, but it's the peer."

Conclusion

Leaving the Appointment

Emma Connell—not her real name—stared at the screen while her stomach twisted: "Application denied—projected lifetime medical costs exceed allowable thresholds." She had played by every rule—taken her medications, attended every appointment, even volunteered for experimental trials. None of it mattered. When she called the insurer, their response was as cold as the machine that made the decision: "The assessment is final." Emma wasn't alone. Across the country, thousands like her were being quietly discarded by a faceless, data-driven system.

This is the dark side of AI. Algorithmic decision-making in healthcare is already shaping who receives coverage. These systems, often shrouded in secrecy, are sold as capable and fair solutions but are too often weaponized against the vulnerable, turning healthcare into a cold calculation of profit over people.

Writer and psychologist Rob Henderson coined the term "luxury beliefs" to refer to the privileged embrace of ideas and concepts by elite groups who never feel the consequences or the effect of their implementation.[1] Such beliefs come with little personal cost, and enhance the status of those who hold them, but can cause serious harm to unseen others.

Luxury beliefs are starkly visible in the intersection of AI and healthcare, epitomized by cases like Emma Connell's. Insurers celebrated AI-driven algorithms as revolutionary tools to optimize costs and streamline decision-making. While executives praised these systems as advancements in efficiency, the burden of rejection fell squarely on individuals like Emma. Such lofty ideals of technological progress often mask a harsher truth whereby the most unwell, impoverished, or marginalized in society pay the highest price.

This book focused on only one justification for technology in healthcare: the physical and psychological constraints associated with traditional primary care consultations. Dissecting the appointment, this is what we found: the visit is riddled with chronic ailments, some of which—like diagnostic errors—the institution of medicine has singularly struggled to come to terms with. Access is much worse for people with greater health needs. Even when we arrive, the format of the traditional visit tilts the balance in favor of physicians, with patients deferring to doctors in ways that stifle honest conversation. This breakdown means some patients are, quite literally, dying of embarrassment.

We learned that medical expertise is a grueling endeavor. Doctors heroically struggle but, because they are only human, unwanted inconsistencies in treatment are inevitable. Worse yet, the very nature of doctors' expertise often blinds them to a spectrum of errors and biases, leaving some of us misdiagnosed, or unfairly treated. This blindness extends to their efforts to eradicate these problems, which are often pitifully weak. And while doctors do their best with bedside manner, they are the worst placed of any health professionals to deliver empathy to patients.

We've considered how technology could address these challenges. While telemedicine—telephone and video consultations—have the potential to expand access to care, especially for the most disadvantaged, it does so imperfectly, and digital divides risk widening health disparity gaps. We discovered that, whereas patients stay schtum with doctors, they pour their hearts out to machines, which can adeptly uncover our most subtle and sensitive signs and symptoms. Although it is early days, there is promising evidence that AI could muffle the noise in medicine, helping to improve consistency in care. Moreover, when carefully designed and implemented, AI has the potential to identify and weed out unwanted biases and reduce the risks of unfair treatment experienced by so many. Chatbots might even do a better job of delivering empathy than doctors, or at least point patients in the direction of real people who can help.

All the appointment's problems are vital issues to address, and technology shows serious promise. But they are by no means the only concerns. To truly advance patient care, this book must be framed within a broader conversation.

Big tech, tough choices, and trust

In an AI-driven world, digital capital—access to internet devices and infrastructure, as well as tech literacy—will increasingly determine the quality of care received. We discovered that these divides are narrowing, but the goalposts of tech are constantly shifting. Large language model AI, and the imminent age of agent AI, introduce new data demands. If current and looming divides are not bridged, this could lead to deeper health disparities. For example, the best diagnostic AI tools may become paywalled with only the wealthiest patients in the richest regions gaining access.

Other pressing considerations demand our deep attention. They include: Who should we trust to run our healthcare systems? Will it be tech giants like OpenAI and Google, whose algorithms are already shaping medical advice and decisions? Who or what should control our health data, and what happens to our privacy in the process? Just how secure is our most sensitive information from exploitation by bad actors? How do we ensure AI systems are reliable for all patients, in all parts of the world, not just a select few? What is the point in diagnostically accurate AI if patients lack basic utilities, electricity, or the medications to treat them? What measures do we use to offset the colossal energy demands on our planet of mass-scale AI adoption in healthcare? And what happens to the people whose jobs go to Dr. Bot?

These urgent questions aren't just about trust, transparency, and the future of healthcare—they're about the kind of society and world we want to live in, one where the line between innovation and exploitation grows ever thinner. The patient stories and advances explored in this book may have sparked these and other concerns in your mind, igniting fresh inquiries and reflections along the way. While I don't claim to address or solve these countless challenges, here I offer some brief reflections on the scale of the problem to offer perspective.

Globally, data breaches are common. By February 2024, of all publicly disclosed attacks, one report estimated that 720 million records, of all kinds, had been compromised worldwide; the actual number is doubtless significantly higher.[2] Healthcare sectors suffer the highest volume of cyber-attacks of all, accounting for around half of all breaches. In the US, the Department of Health and Human Services Office of Civil Rights estimates that, to date, health data breaches have affected one in three Americans.[3]

In Europe the figures are not as damning but hardly glowing. By February 2024, 106 disclosed incidents had affected more than 34 million health records.[2] Although human error is sometimes to blame, most violations involve bad actors. In June 2024, for example, a cyber-criminal gang caused major disruptions to London hospitals, seizing data from 300 million patient interactions from National Health Service (NHS) blood-testing company Synnovis.[4] This was one of the most significant cyber-attacks ever committed in the UK and the criminal gang published private information, including patient names, dates of birth, NHS numbers, and blood test descriptions, on the darknet.

Other data breaches are arguably even more flagrant: they involve the complicity of the very organizations in which we place our trust. In November 2019, for example, *The Wall Street Journal* reported that Ascension, the world's largest Catholic health system and the second-largest "non-profit" health organization in America, collaborated with Google in giving access to the names, dates of birth, and sensitive medical information of 50 million Ascension patients.[5] The collaboration—code-named "Project Nightingale"—involved 150 Google employees; one whistleblower told *The Guardian* that no attempts to de-identify patients' data were made.[6]

The risks do not stop here. Take telemedicine: in the US, in part thanks to the catalyst of COVID-19, it has rapidly become a billion-dollar industry. Many choose on-demand telemedicine for discretion; others see it as a more affordable option than insurance, with its upfront pricing for when they need medical attention. However, patients who choose these consumer telemedicine models face heightened privacy risks. In December 2022, a joint investigation led by journalists Katie Palmer at *STAT* and Todd Feathers and Simon Fondrie-Teitler of *The Markup* revealed companies offering on-demand telemedicine were "leaking sensitive medical information" to the world's largest advertising platforms.[7]

The trio investigated 50 direct-to-consumer telehealth sites and discovered 49 shared health data via Big Tech's tracking tools, which keep records of what websites patients visit, for how long, and what they purchase. In total, they discovered that 13 of the sites embedded at least one tracker from Meta, Google, TikTok, Twitter, Bing, Snap, or LinkedIn. Workit Health, for example, "sent responses about self-harm, drug and alcohol use, and personal information—including first name, email address, and phone number—to Facebook."

Eavesdropping on the minutiae of our daily routines is Big Tech's business model. Hoovering up our daily Google searches, scouring our social media posts, and gleaning our preferences from Amazon purchases, Big Tech companies curate a vast portfolio of data on us.

Smart speakers like Siri and Alexa listen in on our conversations, and chatbots like ChatGPT—and the next wave of "agent AI"—could become even more powerful extractors of our information via their effortless "conversational" abilities.[8]

Harvard Business School professor Shoshana Zuboff dubs this form of profiteering *surveillance capitalism*, whereby masses of information, including our personal details, age, sex, race/ethnicity, financial status, location, spending habits, preferences, health, wellbeing, and how we spend our time are captured and commodified—sold to advertisers and other agents who readily profit from it.[9] As the saying goes, "If you're not paying for the product, you are the product." Like modern-day prospectors in a digital gold rush, Big Tech has proven eager to mine and monetize our personal data derived from our online experiences.

This data can also be used to fuel high-stakes decisions. In the wrong hands, the misuse of sensitive health data carries serious consequences for employment and hiring processes, financial decisions, housing and rental allocation, as well as risk assessments and surveillance in policing and criminal justice.

As consumers and patients, at least at an abstract level, we recognize there is a problem. Recent Pew Research surveys show that around eight in ten Americans believe their personal data is less secure now, and that it is not possible to go through daily life without being tracked.[10] Yet, partly because of legislative apathy, partly because our modern world relies on us being online, and partly because the terms of consent are labyrinthine and unfeasibly onerous, we hand over our information with startling ease.[8]

We are vulnerable online for other—more human—reasons too. When we roam the internet or invite friendly, softly spoken "female" AI agents into our homes, we do not perceive the exploitative nature of surveillance capitalism. This is because we generally don't notice *any* untoward interactions arising at all. Worse still, by creating an illusion of trustworthiness, a range of newer chatbots can even seem kinder, more generous, and more attentive confidants than the flesh-and-blood humans in our lives. AI can supercharge our human sensibilities to trust it. In other words, our inherent biases and blind spots offer Big Tech golden avenues for exploitation.

Why does this happen? Our minds have not evolved to roam around online landscapes, and we are ill-equipped for the stealthy savannahs of cyberspace. We are adapted to understand, and respond to, visible social exchanges and transactions. This makes us especially easy patient-prey for online, for-profit predators. Websites and digital devices create terrains where we are stubbornly oblivious to tracking and tracing. So, while we can entertain the idea of surveillance capitalism at a theoretical level, we fail to grasp on a visceral or intuitive level what is happening to our trail of data. Even when we consciously reflect on the trade-off, at a gut level the "distrust" may feel transient. So, while we can entertain the idea that surveillance capitalism is unfair, or that our privacy is somehow intangibly being threatened, stolen, or exploited, we don't instinctively feel the direct and immediate effects of it.

Without sustained, effortful thinking, our old psychology is simply not attentive or alert to how tech titans monetize our online activities. Again, however, this doesn't let Big Tech off the hook: quite the reverse. We should acknowledge that our evolved, instinctual brainware is fundamentally mismatched for its business model. Our institutions, regulations, and laws must sharpen their teeth, ready to challenge tech when needed. Exploitative or discriminatory actions must face real and meaningful penalties.

Medicine and luxury beliefs

Our doctors work tirelessly, navigating relentless pressures and crumbling health systems, all in a desperate bid to deliver us care. However, throughout this book, we have considered how doctors also unwittingly uphold their own privileged beliefs, which can be at odds with their patients' interests.

While resistance to change in healthcare is understandable and can be well-founded, we found that medicine has a long history of hesitating to embrace groundbreaking innovations, including telemedicine that has improved access to care for so many. While the medical profession—not unreasonably—requests greater resources, and deeper understanding about the challenges faced by its rank and file, too often it embraces or leans into luxury beliefs of its own. The idea that the profession has the wherewithal and the insight to identify its limitations is mistaken. Time and again it relies on the notion that education or training can solve deeply rooted challenges such as discriminatory biases in care, or deficits of empathy, failing to confront the scant evidence for these claims for the very people who matter—patients.

Even when contemplating change and embracing AI, medicine assumes that teamwork with technology will provide a straightforward remedy, conjuring up soft-focus images of "man and machine." As we've learned, these are medical mirages. Implementing these tools is an ongoing challenge.[11] Moreover, probing the psychological dynamics of the visit, there are no guarantees that doctors and AI will metaphorically walk off, hand in hand, into the sunset, securing a happy-ever-after healthcare.

Doctors face immense challenges in their work, but they are, above all, human. The profession's actions often stem from profoundly understandable inclinations to defend their authority. Confronted by an identity crisis, many physicians will seek to conserve their professional identity. They may justifiably be concerned about AI's threats to their societal status, earnings, and self-worth.

American author Clay Shirky, known for his work on the internet's societal impact, coined the now-famous "Shirky Principle," stating: "Institutions will try to preserve the problem to which they are the solution."[12] In the case of medicine, I argue that the situation is even worse. Faced with increasingly broken health systems, the profession is failing to constructively reimagine and work toward

CONCLUSION

credible solutions. This is a claim that merits support. So, as we transition from the traditional appointment, it is worth taking a moment to reflect on how deeply occupational allegiances have shaped doctors' responses to AI.

Data entry and paperwork have long plagued healthcare professionals, and doctors are eager to automate drudgery. Before the phrase "large language models" gained prominence, or anyone had heard of OpenAI, in early surveys that I conducted doctors decisively singled out documentation as the only medical task they were certain AI would revolutionize.[13–18] They were fiercely confident about it happening fast. It is therefore no surprise they are enthusiastic about adopting AI tools that can help them.[19,20] Recently, Microsoft collaborated with Epic Systems—a leading US software company with an estimated 78 percent share of hospital electronic medical record systems in the United States—through its Azure platform, to offer a privacy-compliant AI service to support healthcare documentation.[21]

Voice-to-text clinical note generation products also represent a growing space in healthcare: apps such as Ambient Experience from Nuance can listen to physicians' and patients' conversations and create clinical notes in real time, ready for review. As we discovered, these tools are far from perfect, although in the US and Australia such capacities are rapidly becoming embedded into electronic health systems, signaling radical changes to medical practices.

Physicians' forecasts on documentation have begun to materialize. But this was not because, figuratively speaking, they had their fingers on the pulse of bots. Rather, my research identified a lack of equanimity in doctors' predictions about technology,[14,15] reflecting wishful thinking. Some doctors were even prescriptive about it; one British GP told us, "Please hurry up with the technological advances to take away some of the crap that I still have to sort out—then I will be able to get back to proper diagnosing and doctoring."[14]

Resistance to AI across the board in healthcare—aside from its use in documentation—stems from legacy beliefs, deeply rooted assumptions about how medicine ought to be practiced. Such rigid perspectives ignore the serious scope for AI. Doctors, we might argue, should not get to cherry-pick the aspects of their work they want to preserve on the grounds of prestige, professional meaning, or cultural antecedent. Overlooking AI's potential and neglecting to harness its benefits—or to prepare for its risks—leads to critical missteps that could once again hinder progress and delay the transformative advances healthcare desperately needs. Disdain, tepid engagement, or amateur interest won't suffice. Given the challenges facing the traditional appointment and the potential of AI, medicine's response, in the words of historian David Wootton, risks "postponing progress."[22]

The fact is, technology is not only coming for paperwork, which doctors justifiably consider well beneath them. Just as in other professions, it is coming for other aspects of their job too.

In 2019, my team polled an international panel of leading AI healthcare experts. We asked their opinions on how AI would change American primary care by 2029.[23] Their forecasts were modest and, among their key predictions, experts anticipated technology would lead to greater access to care and increased diagnostic accuracy, especially among those with limited access to doctors, that is, minorities and people with rare diseases. They also anticipated workplace changes: incursions of AI into physician expertise, increased AI training requirements for medical students, and more students with engineering and computing backgrounds entering the profession.

A return to the traditional status quo is the least likely trajectory. Yet over the past seven years, multiple surveys that I have conducted demonstrate that most doctors are profoundly ill-prepared for the digital age.[13,14,17,24] Despite small pockets of activity in elite teaching hospitals, the picture is dispiriting: AI education is not routine.[17,25]

Medical education is stuck in a twentieth-century rut, and most medical programs are still training students to be analogue doctors. In March 2024, in the very latest curricular specifications described by the Liaison Committee on Medical Education—the body that outlines the standards for medical degrees awarded in Canada and the United States—education about AI is missing.[26] Despite recommendations for digital literacy and medical curricular reform in the UK laid out in 2019 in the Topol Review, the pace of educational change is glacial.[27] There is still a failure to strengthen even minimal levels of digital literacy in British medical schools: in its "outcomes for graduates" guidance, for example, the General Medical Council omits any mention of the words "artificial intelligence" or "machine learning."[28]

The profession needs to wake up and confront the reality of algorithm-driven healthcare. In the coming years, we can expect medicine to enter a state of flux as AI continues to automate, and innovate, doctors' tasks. This makes it more critical than ever for the medical community to educate doctors about digital tools and their transformative potential. They need to know about the problems for which technology could present solutions. They also need to know the current limits of these tools, and what they can't yet do.

From AI hallucinations to unwanted biases, and from too much algorithmic appreciation to too much aversion, doctors need to stay sharp to the challenges—to understand the potential of AI but also see its limitations. Instead of turning a blind eye to bots, medicine must cultivate clinicians with critical aptitude. A new generation of clinicians needs to know when and why AI tools are genuinely helpful and when they might cause real harm. Quite simply, medical schools must do a better job of engaging in the modern machine era.

We should therefore expect clinical education to evolve and change. Do we really need future physicians who can memorize the Krebs cycle as they always have? Should we still encourage them to emulate senior doctors who have spent their careers shunning technology? Or should we seek to carve out new clinician relationships with AI for the twenty-first century? It is not only *what* we teach that is relevant but *who* teaches *whom*.

Do we need a medical profession?

Traditionally, we have depended on a "grand bargain," where patients trust medical professionals to be accessible, affordable, and to keep up to date, applying their expertise to meet our clinical needs.[29] However, this book has critically examined the performance of doctors at each stage of the traditional medical appointment. We've examined why the grand bargain can fail patients (and doctors) at each phase—for very *human* reasons. We've emphasized throughout that the function of medicine is the reliable care of the patient: with this in mind, we have also asked whether technology, or other kinds of workers, could assist with key tasks, or even replace doctors (*Homo medicus*). In summary, we discovered that there may be excellent psychological justifications for realigning responsibilities in healthcare.

This sparks essential questions about which roles might shift and whether—in the long term—the traditional medical profession is even necessary. Here are some ideas: chatbots could increasingly triage our primary care, referring us to doctors when we most need them. When we see doctors, their role may shift to becoming skilled interpreters of AI-driven advice. Or it could even be that an expanding profession of physician assistants, nurse practitioners, or more junior physicians would be just as well or better suited to team up with chatbots in critiquing AI's clinical judgments. Alternatively, a new class of coworkers— "clinical AI specialists"—could be trained in medicine and in analyzing the outputs of diagnostic tools in ways that maximize critical engagement and minimize health expert hubris. In the long term such arrangements could even give way to AI agents critiquing each other.

These are far from the only careers or post-physician roles we can envisage. If and when white coats and stethoscopes are gradually, then finally, hung up, most collars in the future of medicine will be hoodies. If we choose to, we could build an ethical, technologically transformed medicine which fragments and proliferates into an abundance of new posts.

CONCLUSION

Here are a few: AI-powered telemedicine developers could explore how platforms can enhance remote patient monitoring and video visits. Medical data scientists could specialize in curating and analyzing large sets of data to extract meaningful medical insights that can improve patient care. Predictive analytics experts could develop models to forecast patient outcomes, prognoses, and drug repurposing.

Medical knowledge engineers, meanwhile, could train and test AI models and investigate the quality of the data fed to AI. They could assess which inputs or prompts elicit optimal responses from bots, and examine how to safely solicit patient information, triage illnesses, and undertake diagnostics. Working alongside them, ethically literate health engineers could work to eliminate unwanted biases in medical decisions and assess the safety, accuracy, and fairness of these tools for different patient populations. Their colleagues, the AI clinical research coordinators, could manage clinical trials and other studies investigating the effectiveness of digital applications in clinical settings.

Meanwhile at the user-end, AI healthcare engineers could examine how these tools can be integrated into existing healthcare environments, including electronic health records. Research and design professionals could craft chatbot interfaces and avatars that encourage patient users to open up, seamlessly guiding them to share symptoms while striking a psychological balance that fosters justified trust and meaningful engagement.

At a health-system level, there could be scalable oversight too. AI could be designed to oversee every phase of the clinical process and to monitor for biases, disparities, and aberrations in care. Equally and reciprocally, AI process managers could oversee the AI, scrutinizing and supervising the learning that is gleaned from red flags and outcome observations. If there is a tendency to over-diagnose and over-treat, both the AI and the process managers will spot this. AI medical ethics officers could ensure technologies focus on safety, patient privacy, and regulatory compliance, and avoid algorithmic discrimination. At every stage, patients could play a routine, consultative, and anchoring role, ensuring the voice of the service user is not overshadowed, ignored, or overlooked.

Nor need all the roles be technology related. Coaches, and peer supporters, could be some of the best-placed humans to assist us when we need a literal shoulder to cry on, or to help us get a healthy routine on track. Digital health navigators could guide us through the complexities of modern health technologies—whether it's downloading apps, providing advice on digital privacy, or endorsing the safest and most effective tools with trusted certifications. Even if we do still desire the single figurehead of a healer—a "doctor"—we can guarantee that teams of people and technology will be working behind the scenes.

Should we decide that this is the direction we want for healthcare, the opportunities could grow. Indeed, it might surprise you to hear that many of these roles already exist. They are not yet the stuff of TV medical dramas, memoirs, or film but together they embody a new breed of clinical heroes.

Professor Matt Might and his undergraduate students are already worthy candidates. Earlier in the book, we met the trailblazing computer scientist and rare disease warrior who uncovered his son Bertrand's killer. Professor Might's dedication to unlocking the secrets within clinical data has already paved the way for new roles.

As part of the Biomedical Data Translator Project at the National Institutes of Health, Professor Might contributes his expertise as a key member of the consortium. He told me that the project comprises two kinds of teams: "There's knowledge collection teams, and they're trying to vacuum up biomedical knowledge. And then there's the reasoning teams like us, where we take that knowledge and then make inferences on top of it, looking for the hidden connections that have presumably always been there, in the data, or in the literature."

The project has been ongoing for around eight years. At the University of Birmingham, Alabama, Matt's undergraduate students are trained to use an AI tool called MediKanren. The word "*kanren*" comes from the Japanese word "関連" which means "connection."

MediKanren is an AI system that can reason over and make connections between different pieces of biomedical knowledge. It deciphers basic scientific biomedical research—like X inhibiting Y and Y causing Z—and seamlessly connects these conclusions, making it a powerful silicon Sherlock Holmes of healthcare.

Students process data about rare diseases, running queries and acting as interpreters. They determine which queries to run and curate the results. "To the outside world," he says, "it makes the undergraduates look like superhumans." As a result of this process, their research reports are carefully structured to contain no explicit medical advice.

However, Professor Might says, "There's a lot of: 'if you connect the dots, there's very useful information here'. But it has to be the treating physician on the other side who connects the dots." He adds, "It's like setting up the tee ball. *Good luck missing*." In six years, Professor Might's team at the University of Alabama has engaged with 600 patients, providing medically beneficial treatment suggestions to nearly half.

Echoing Kasparov's Law from the last chapter—Weak human + machine + better process—and much like the two amateurs in the "ZackS" chess team who bested the formidable grandmasters, Matt Might trained undergraduates working with MediKanren to outshine eminent doctors. His university doesn't promote these research services—word of mouth alone keeps the team in constant demand. Regrettably, medicine has yet to unlock the full potential of these transformative partnerships.

Tomorrow's patients

Throughout history, medicine has considered us—patients—as the passive, unenlightened recipients of its expertise. In the twenty-first century, this no longer cuts it. Patients are increasingly taking control of their care, and there's no going back.

Liam from Boston, whose atrial fibrillation was diagnosed by his Apple Watch, told me that chatbots filled in the gaps in communication omitted by his cardiologist. "I asked it [ChatGPT] about the calcification of the aorta, and it gave me accurate descriptions of it."

"I've looked up drugs, the names of drugs, because I'm on five different medications. I've checked up on them, how good they are, how widespread they are. It gave me the time that my doctor didn't."

Jen Lawson, who lives with Ehlers-Danlos syndrome, uses them too. "Bots help me talk to my doctor," she admits. "I tell the bot, 'This is what I want to say, how do I make it less spicy? Make it forceful but formal.'"

Behaving like a cyber-Cyrano de Bergerac, the chatbot is now part of Jen's routine clinical correspondence. "It helps me get rid of the 'Jen voice'. Doctors take me more seriously. I have the bot to thank for that."

Jen's father, Tom Lawson, the 93-year-old academic who underwent knee surgery, told me that he too checks what his doctor is telling him. "I would second-guess the doctor *anytime*. By the way," Tom adds, emphatically, "I *always* double-check what the AI is saying too."

This back-and-forth checking—acting as both a first and second layer of validation both for doctors' opinions and AI output—could become a subtle yet growing trend.

Dr. Adam Rodman's experiences resonate with these perspectives too. An internist at Beth Israel Medical Center, Assistant Professor at Harvard Medical School, and a pioneer of research into large language model (LLM) bots in medicine, Adam openly revealed to me that, as a parent, he has found these tools empowering.

"My kid has eye problems and has had multiple eye surgeries. We thought he had a brain tumor. He doesn't. But optometry notes are written ... well it might as well be hieroglyphics. As an internist," Adam asserts, "I know what I know, and I know what I don't know."

"So, my wife and I took one of his notes in and asked ChatGPT to interpret it into regular language. And it did. It perfectly explained what was going on."

CONCLUSION

Those who recall the early days of the internet will remember: it took years for consumers to grasp its full potential. LLM tools are in a similar phase—widespread adoption has yet to reach its tipping point, but the momentum is unmistakable. Just like Liam, Jen, Tom, and Adam, by August 2024, a health-tracking poll in the US reported that around one in six adults (17 percent) used AI chatbots at least once a month for health information and advice; this figure rose to one quarter of adults aged 30 and younger.[30]

Beyond patient empowerment via chatbots, more advanced AI agents could one day become our personalized clinicians. Designed to match our preferences and tailored to our unique medical needs, bespoke bot companions might eventually revolutionize healthcare for good. They could talk to us in real time, offer tips for illness prevention, give personalized advice on health risks informed by our medical history, and deliver diagnostic insights with unparalleled precision and compassion.

Similarly, in the future, bots might not only provide first or second opinions; they might challenge and refine the conclusions of other bots. As AI progresses, human doctors working alongside technology could become the riskiest form of care, leaving the best AI-driven medicine accessible only to those who can afford it. Alternatively, or in parallel, those in resource-limited countries might leapfrog to better medical care by sidestepping the tangled inefficiencies and traditions embedded in august, time-honored human healthcare systems and their revered institutions.

One thing is clear: just as there is no pill for every ill, no technology will be a panacea for medicine's problems. All innovations invite a host of both expected and unanticipated problems, and these must be loudly and openly discussed. Such debates should be measured, honest, and fearless. To this, I add: numerous pioneering AI researchers and journalists have made invaluable contributions in exposing flaws, monitoring progress, and holding current technologies to account.

Yet commentators too can succumb to luxury beliefs—pursuing narrow agendas, ignoring the bigger picture, or presuming to speak for voiceless patients without genuinely understanding their diversity or the variety of their needs. Insulated by privileged and powerful positions, their criticisms can sound virtuous yet carry little personal cost. However, inaction is a choice too. The conservative dependency on inherited legacy structures in healthcare also carries moral consequences. From the shelter of exclusive circles, it is easy to advocate against innovation without experiencing the real-world consequences of this choice that less privileged individuals face every day.

The perfect is the enemy of the good. Yet too often, debates around AI default to contrasting its performance against the idealized luxury of the very best doctors, setting an impractical and skewed benchmark. This is not the world most of us live in. The correct comparison for the global majority is not with AI versus prompt medical attention, it is with getting no medical attention at all. Again, even this consideration should not overshadow wider debates. The risks and benefits of what technology can offer, and how it can be tamed, will need to be pursued actively and robustly, with moral imagination.

This book recognizes the debt of gratitude that we owe doctors. They work tirelessly in our service, confronting demands that other professions simply don't face. Doctors have made, and continue to make, profound personal sacrifices in the line of duty.

However, it is patients who must get the final word. If we could start from scratch, I wager, we would rationally choose to remodel medicine. An upgraded healthcare system could allow many more of us—whoever we are, whatever our problems, and wherever we live—to access reliable expertise more quickly and affordably. If we technologically revamped healthcare, it could mean the function of medicine was working—perhaps not perfectly, but better than what we've got. For patients and their families, this might be all they truly need—or ever wanted.

Notes

Introduction: The Ailing Appointment

1. https://www.bmj.com/content/353/bmj.i2139/ (2016).
2. Singh H et al. The frequency of diagnostic errors in outpatient care: estimations from three large observational studies involving US adult populations. *BMJ Qual Saf* 2014; 23(9): 727–731.
3. Newman-Toker DE et al. Burden of serious harms from diagnostic error in the USA. *BMJ Qual Saf* 2024; 33: 109–120.
4. Cheraghi-Sohi S et al. Incidence, origins and avoidable harm of missed opportunities in diagnosis: longitudinal patient record review in 21 English general practices. *BMJ Qual Saf* 2021; 30: 977–985.
5. Graber ML. The incidence of diagnostic error in medicine. *BMJ Qual Saf* 2013; 22: ii21–ii27.
6. Shojania KG et al. Changes in rates of autopsy-detected diagnostic errors over time: a systematic review. *JAMA* 2003 Jun 4; 289(21): 2849–2856.
7. Singh H et al. The global burden of diagnostic errors in primary care. *BMJ Qual Saf* 2017; 26: 484–494.
8. https://cordis.europa.eu/project/id/830017 (Aug 23, 2022).
9. Newman-Toker DE et al. Serious misdiagnosis-related harms in malpractice claims: the "Big Three"—vascular events, infections, and cancers. *Diagnosis* 2019; 6: 227–240.
10. https://www.medscape.com/slideshow/2024-lifestyle-burnout-6016865 (2024).
11. https://www.bma.org.uk/news-and-opinion/stress-on-doctors-reaches-all-time-high (2023).
12. Prasad K et al. Prevalence and correlates of stress and burnout among US healthcare workers during the COVID-19 pandemic: a national cross-sectional survey study. *EClinicalMedicine* 2021; 35: 100879.
13. Khan N et al. Cross-sectional survey on physician burnout during the COVID-19 pandemic in Vancouver, Canada: the role of gender, ethnicity and sexual orientation. *BMJ Open* 2021; 11: e050380.

14. https://www.independent.co.uk/news/health/coronavirus-burnout-nhs-doctors-nurses-second-wave-cases-deaths-latest-b1256453.html (Mar 8, 2021).
15. Olfson M et al. Suicide risks of health care workers in the US. *JAMA* 2023 Sep 26; 330(12): 1161–1166.
16. https://www.theguardian.com/us-news/2023/sep/26/surgeons-suicide-doctors-physicians-mental-health (Sept 26, 2023).
17. Dyrbye LN et al. Medical licensure questions and physician reluctance to seek care for mental health conditions. *Mayo Clinic Proceedings*, 2017; 92(10): 1486–1493.
18. https://www.who.int/news-room/fact-sheets/detail/ageing-and-health (Oct 1, 2024).
19. https://www.un.org/en/dayof8billion (Nov 15, 2022).
20. Ansah JP, Chiu C-T. Projecting the chronic disease burden among the adult population in the United States using a multi-state population model. *Front Public Health* 2023; 10: 1082183.
21. https://www.nhs.uk/news/neurology/dementia-could-strike-1-in-3-born-this-year-claims-report/ (2015).
22. https://time.com/6199666/physician-shortage-challenges-solutions/ (July 25, 2022).
23. https://www.politico.eu/article/france-doctors-europe-too-far-too-old-too-few/ (2022).
24. https://sante.gouv.fr/actualites/presse/communiques-de-presse/article/une-avancee-majeure-pour-le-systeme-de-sante-l-ensemble-des-ordres-des (Oct 13, 2022).
25. https://www.independent-practitioner-today.co.uk/2022/09/half-of-trainees-want-to-quit-job/ (2022).
26. Tawfik DS et al. Physician burnout, well-being, and work unit safety grades in relationship to reported medical errors. *Mayo Clinic Proceedings*, 2018; 93(11): 1571–1580.
27. Shanafelt TD et al. Longitudinal study evaluating the association between physician burnout and changes in professional work effort. *Mayo Clinic Proceedings*, 2016; 91(4): 422–431.
28. Croskerry P. Achieving quality in clinical decision making: cognitive strategies and detection of bias. *Acad Emerg Med* 2002; 9: 1184–1204.
29. Shanafelt TD et al. Burnout and self-reported patient care in an internal medicine residency program. *Ann Intern Med* 2002; 136: 358–367.
30. Pereira-Lima K et al. Association between physician depressive symptoms and medical errors: a systematic review and meta-analysis. *JAMA Netw Open* 2019; 2: e1916097–e1916097.
31. Fahrenkopf AM et al. Rates of medication errors among depressed and burnt out residents: prospective cohort study. *BMJ* 2008; 336: 488–491.
32. Caruso R et al. Violence against physicians in the workplace: trends, causes, consequences, and strategies for intervention. *Curr Psychiatry Rep* 2022 Dec; 24(12): 911–924.
33. Nunez-Smith M et al. Health care workplace discrimination and physician turnover. *J Natl Med Assoc* 2009; 101: 1274–1282.
34. https://www.england.nhs.uk/2020/02/nhs-staff-morale-improves-but-too-many-facing-abuse/ (2020).

35. Studdert DM et al. Defensive medicine among high-risk specialist physicians in a volatile malpractice environment. *JAMA* 2005 Jun 1; 293(21): 2609–2617.
36. De González AB et al. Projected cancer risks from computed tomographic scans performed in the United States in 2007. *Arch Intern Med* 2009; 169: 2071–2077.
37. https://www.health.harvard.edu/cancer/radiation-risk-from-medical-imaging (2021).
38. Sodickson A et al. Recurrent CT, cumulative radiation exposure, and associated radiation-induced cancer risks from CT of adults. *Radiology* 2009; 251: 175–184.
39. Localio AR et al. Relation between malpractice claims and adverse events due to negligence: results of the Harvard Medical Practice Study III. *N Engl J Med* 1991; 325: 245–251.
40. https://www.ama-assn.org/practice-management/sustainability/1-3-physicians-has-been-sued-age-55-1-2-hit-suit (2018).
41. Wallace E et al. The epidemiology of malpractice claims in primary care: a systematic review. *BMJ Open* 2013 Jun 1; 3(7): e002929.
42. https://www.ft.com/content/30677465-33bb-4f74-a8e6-239980091f7a (Oct 1, 2024).
43. Sellars W. Philosophy and the scientific image of man. *Front Sci Philos* 1962; 1: 35–78.

1 Prognosis and Treatment Options

1. Rosenbaum EE. *A Taste of My Own Medicine: When the Doctor is the Patient.* Random House New York, 1988.
2. Meyer AN et al. Physicians' diagnostic accuracy, confidence, and resource requests: a vignette study. *JAMA Intern Med* 2013; 173: 1952–1958.
3. Ludikhuize J et al. How nurses and physicians judge their own quality of care for deteriorating patients on medical wards: self-assessment of quality of care is suboptimal. *Crit Care Med* 2012; 40: 2982–2986.
4. Podbregar M et al. Should we confirm our clinical diagnostic certainty by autopsies? *Intensive Care Med* 2001; 27: 1750–1755.
5. Trivers R. *The Folly of Fools: The Logic of Deceit and Self-deception in Human Life*. Basic Books, 2011.
6. Trivers R. The elements of a scientific theory of self-deception. *Ann N Y Acad Sci* 2000; 907: 114–131.
7. Cavalcanti RB, Sibbald M. Am I right when I am sure? Data consistency influences the relationship between diagnostic accuracy and certainty. *Acad Med* 2014; 89: 107–113.
8. Lowry F. Failure to perform autopsies means some MDs "walking in a fog of misplaced optimism." *CMAJ Can Med Assoc J* 1995; 153: 811.
9. Voltaire F. *Candide, or Optimism*. Penguin UK, 2013.
10. Donaldson MS et al. *To Err Is Human: Building a Safer Health System*. National Academies Press, 2000.
11. Robinson AR et al. Physician and public opinions on quality of health care and the problem of medical errors. *Arch Intern Med* 2002; 162: 2186–2190.
12. Newman-Toker DE, Pronovost PJ. Diagnostic errors—the next frontier for patient safety. *JAMA* 2009; 301: 1060–1062.

13. Wachter RM. Why diagnostic errors don't get any respect—and what can be done about them. *Health Aff (Millwood)* 2010; 29: 1605–1610.
14. https://www.improvediagnosis.org/ (2024).
15. Balogh E et al. (eds). *Improving Diagnosis in Health Care*. National Academies Press, 2015.
16. Wilson TD. *Strangers to Ourselves*. Harvard University Press, 2004.
17. Hayward RA, Hofer TP. Estimating hospital deaths due to medical errors: preventability is in the eye of the reviewer. *JAMA* 2001; 286: 415–420.
18. McDonald CJ et al. Deaths due to medical errors are exaggerated in Institute of Medicine report. *JAMA* 2000; 284: 93–95.
19. Gianoli GJ. Medical error epidemic hysteria. *Am J Med* 2016; 129: 1239–1240.
20. Croskerry P et al. Emotional influences in patient safety. *J Patient Saf* 2010; 6: 199–205.
21. Blendon RJ et al. Views of practicing physicians and the public on medical errors. *N Engl J Med* 2002; 347: 1933–1940.
22. Kay A. *This Is Going to Hurt: Secret Diaries of a Junior Doctor*. Pan Macmillan, 2017.
23. Marsh H. *Do No Harm: Stories of Life, Death and Brain Surgery*. Hachette UK, 2014.
24. Wootton D. *Bad Medicine: Doctors Doing Harm since Hippocrates*. Oxford University Press, 2007.
25. Kuhn TS. *The Structure of Scientific Revolutions*. University of Chicago Press, 1962.
26. https://blogs.bmj.com/bmj/2010/09/22/richard-smith-computers-take-histories-better-than-doctors-why-dont-they-do-it-more/ (2010).
27. https://www.washingtonpost.com/news/to-your-health/wp/2015/11/03/want-to-reach-your-doc-many-americans-would-like-to-use-email-or-text/ (Nov 3, 2015).
28. https://news.bloomberglaw.com/health-law-and-business/health-care-clings-to-faxes-as-u-s-pushes-electronic-records (2021).
29. https://www.bbc.com/news/health-47215799 (2019).
30. https://www.thetimes.co.uk/article/how-to-fix-the-nhs-times-health-commission-qpdgfwzvg (Jul 2, 2023).
31. Ofri D. *What Doctors Feel: How Emotions Affect the Practice of Medicine*. Beacon Press, 2013.
32. Singh H et al. Measures to improve diagnostic safety in clinical practice. *J Patient Saf* 2019 Dec 1; 15(4): 311–316.
33. Susskind RE, Susskind D. *The Future of the Professions: How Technology Will Transform the Work of Human Experts*. Oxford University Press, 2015.
34. Freidson E. *Professional Dominance: The Social Structure of Medical Care*. Routledge, 2017.
35. Blease C et al. Patients, clinicians and open notes: information blocking as a case of epistemic injustice. *J Med Ethics* 2022 Oct 1; 48(10): 785–793.
36. Hägglund M et al. Patient empowerment through online access to health records. *BMJ* 2022 Sep 29; 378: e071531.
37. Walker J et al. OpenNotes after 7 years: patient experiences with ongoing access to their clinicians' outpatient visit notes. *J Med Internet Res* 2019; 21: e13876.

38. Bell SK et al. Tackling ambulatory safety risks through patient engagement: what 10,000 patients and families say about safety-related knowledge, behaviors, and attitudes after reading visit notes. *J Patient Saf* 2021; 17: e791–e799.
39. Bell SK et al. Frequency and types of patient-reported errors in electronic health record ambulatory care notes. *JAMA Netw Open* 2020; 3: e205867–e205867.
40. Blease CR et al. Experiences and opinions of general practitioners with patient online record access: an online survey in England. *BMJ Open* 2024; 14: e078158.
41. https://www.bma.org.uk/pay-and-contracts/contracts/gp-contract/gp-contract-changes-england-202324 (2023, accessed May 15, 2024).
42. https://www.opensecrets.org/federal-lobbying/top-spenders (2024).
43. https://www.technologyreview.com/2025/01/21/1110260/openai-ups-its-lobbying-efforts-nearly-seven-fold/ (2025).
44. Hooker RS et al. Career flexibility of physician assistants and the potential for more primary care. *Health Aff (Millwood)* 2010; 29: 880–886.
45. Swan M et al. Quality of primary care by advanced practice nurses: a systematic review. *Int J Qual Health Care* 2015; 27: 396–404.
46. Laurant M et al. Substitution of doctors by nurses in primary care. *Cochrane Database of Systematic Reviews* 2005(2).
47. https://www.statnews.com/2018/02/21/health-providers-shortage/ (2018).
48. https://www.clinicaladvisor.com/home/topics/practice-management-information-center/ama-opposes-full-potential-nps-pas/ (2023).
49. https://www.theguardian.com/society/2024/jan/18/physician-associates-arent-doctors-and-shouldnt-be-regulated-as-such-says-bma (Jan 18, 2024).
50. https://www.gao.gov/assets/gao-18-88.pdf (2017).
51. DuBois JM et al. Serious ethical violations in medicine: a statistical and ethical analysis of 280 cases in the United States from 2008–2016. *Am J Bioeth* 2019; 19: 16–34.
52. Harris JA, Byhoff E. Variations by state in physician disciplinary actions by US medical licensure boards. *BMJ Qual Saf* 2017; 26: 200–208.
53. https://www.fsmb.org/advocacy/news-releases/national-survey-indicates-majority-of-physician-misconduct-goes-unreported/ (2019).
54. Campbell EG et al. Professionalism in medicine: results of a national survey of physicians. *Ann Intern Med* 2007; 147: 795–802.
55. Gallagher TH et al. Talking with patients about other clinicians' errors. *N Engl J Med* 2013; 369: 1752–1757.
56. Schwartz WB. Medicine and the computer: the promise and problems of change. In: Anderson JG, Jay SJ (eds) *Use and Impact of Computers in Clinical Medicine*. Springer New York, 1970, pp. 321–335.
57. Quoted in https://www.kqed.org/futureofyou/274449/will-computers-ever-be-able-to-make-diagnoses-as-well-as-physicians (2016).
58. https://www.theguardian.com/technology/2014/jun/15/robot-doctors-online-lawyers-automated-architects-future-professions-jobs-technology (Jun 15, 2014).
59. https://www.zdnet.com/article/qa-andrew-mcafee-erik-brynjolfsson-co-authors-of-the-second-machine-age/ (2014).
60. https://www.wired.co.uk/article/ibm-watson-medical-doctor (2013).

61. https://www.youtube.com/watch?v=2HMPRXstSvQ (2016).
62. https://www.statnews.com/2023/05/03/artificial-intelligence-doctor-vinod-khosla-ventures/ (2023).
63. https://www.nytimes.com/2023/01/06/podcasts/transcript-ezra-klein-interviews-gary-marcus.html (Jan 6, 2023).
64. https://www.theglobeandmail.com/opinion/article-just-what-the-doctor-ordered-how-ai-will-change-medicine-in-the-2020s/ (Dec 27, 2019).
65. Christensen CM. *The Innovator's Dilemma: When New Technologies Cause Great Firms to Fail*. Harvard Business Review Press, 2013.
66. https://hbr.org/2015/12/what-is-disruptive-innovation (2015).
67. Bacon F. *Francis Bacon: The New Organon*. Cambridge University Press, 2000.
68. https://www.prnewswire.com/news-releases/artificial-intelligence-ai-in-healthcare-market-worth-148-4-billion-by-2029---exclusive-report-by-marketsandmarkets-302052956.html (2024).
69. Blease C et al. Psychiatrists' experiences and opinions of generative artificial intelligence in mental healthcare: an online mixed methods survey. *Psychiatry Res* 2024; 333: 115724.
70. Blease C et al. Generative artificial intelligence in primary care: an online survey of UK General Practitioners. *BMJ Health Care Inform* 2024 Aug 29; 31: e101102.

2 Patient Pilgrims

1. https://www.bls.gov/regions/midwest/summary/blssummary_kalamazoo.pdf (2024).
2. https://www.who.int/news-room/fact-sheets/detail/falls (2021).
3. Brophy M et al. Injuries among US adults with disabilities. *Epidemiology* 2008; 465–471.
4. https://www.who.int/news/item/13-12-2017-world-bank-and-who-half-the-world-lacks-access-to-essential-health-services-100-million-still-pushed-into-extreme-poverty-because-of-health-expenses (2017).
5. Moreno-Serra R, Smith PC. Does progress towards universal health coverage improve population health? *The Lancet* 2012; 380: 917–923.
6. https://www.thebalance.com/universal-health-care-4156211 (2020).
7. https://www.cdc.gov/nchs/data/nhsr/nhsr169.pdf (2022).
8. Woolhandler S, Himmelstein DU. The relationship of health insurance and mortality: is lack of insurance deadly? *Ann Intern Med* 2017; 167: 424–431.
9. Himmelstein D et al. *Medical Bankruptcy: Still Common Despite the Affordable Care Act*. American Public Health Association, 2019.
10. Blumenthal D et al. Covid-19—implications for the health care system. *N Engl J Med* 2020 Oct 8; 383: 1483–1488.
11. Galvani AP et al. Universal healthcare as pandemic preparedness: the lives and costs that could have been saved during the COVID-19 pandemic. *Proc Natl Acad Sci* 2022; 119: e2200536119.
12. https://ec.europa.eu/health/system/files/2020-12/2020_healthatglance_rep_en_0.pdf (2020).
13. https://www.who.int/news-room/fact-sheets/detail/disability-and-health (2018).

14. https://www.bls.gov/news.release/disabl.nr0.htm (2024).
15. Krahn GL et al. Persons with disabilities as an unrecognized health disparity population. *Am J Public Health* 2015; 105: S198–S206.
16. Mitra S et al. Extra costs of living with a disability: a review and agenda for research. *Disabil Health J* 2017; 10: 475–484.
17. Marmot MG. Status syndrome: a challenge to medicine. *JAMA* 2006; 295: 1304–1307.
18. https://www.aha.org/system/files/2019-02/rural-report-2019.pdf (2019).
19. Scotti S. Tracking rural hospital closures. *NCSL legisbrief* 2017 Jun 1; 25(21): 1–2.
20. https://ec.europa.eu/eurostat/statistics-explained/index.php?title=Urban-rural_Europe_-_quality_of_life_in_rural_areas#Health (2023).
21. Kullgren JT et al. Nonfinancial barriers and access to care for US adults. *Health Serv Res* 2012; 47: 462–485.
22. Long SK, Phadera L. Barriers to obtaining health care among insured Massachusetts residents. *Health Aff (Millwood)* 2019; 38(1): 52–58.
23. https://www.brinknews.com/quick-take/the-us-has-one-of-the-worst-public-transit-systems-in-the-world/ (2022).
24. Syed ST et al. Traveling towards disease: transportation barriers to health care access. *J Community Health* 2013; 38: 976–993.
25. Silver D et al. Transportation to clinic: findings from a pilot clinic-based survey of low-income suburbanites. *J Immigr Minor Health* 2012; 14: 350–355.
26. Sakellariou D, Rotarou ES. Access to healthcare for men and women with disabilities in the UK: secondary analysis of cross-sectional data. *BMJ Open* 2017; 7: e016614.
27. https://www.eesc.europa.eu/en/news-media/news/denied-right-health-persons-disabilities-have-more-difficulty-accessing-healthcare (2023).
28. Ray KN et al. Disparities in time spent seeking medical care in the United States. *JAMA Intern Med* 2015; 175: 1983–1986.
29. Mechakra-Tahiri SD et al. The gender gap in mobility: a global cross-sectional study. *BMC Public Health* 2012; 12: 598.
30. Lee A et al. Assessment of parking fees at National Cancer Institute–designated cancer treatment centers. *JAMA Oncol* 2020 Aug 1; 6(8): 1295–1297.
31. https://www.cdc.gov/disability-and-health/media/pdfs/disability-impacts-all-of-us-infographic.pdf (2024).
32. Freiberger E et al. Mobility in older community-dwelling persons: a narrative review. *Front Physiol* 2020; 11: 881.
33. Oostrom T et al. Outpatient office wait times and quality of care for Medicaid patients. *Health Aff (Millwood)* 2017; 36: 826–832.
34. Kirschner KL et al. Structural impairments that limit access to health care for patients with disabilities. *JAMA* 2007; 297: 1121–1125.
35. Lagu T et al. Access to subspecialty care for patients with mobility impairment: a survey. *Ann Intern Med* 2013; 158: 441–446.
36. Iezzoni LI et al. Physicians' perceptions of people with disability and their health care: study reports the results of a survey of physicians' perceptions of people with disability. *Health Aff (Millwood)* 2021; 40: 297–306.
37. Sakellariou D et al. Barriers to accessing cancer services for adults with physical disabilities in England and Wales: an interview-based study. *BMJ Open* 2019; 9: e027555.

38. Hart JT. The inverse care law. *The Lancet* 1971; 297: 405–412.
39. https://ncd.gov/newsroom/2021/federal-report-release-covid-19 (2021).
40. Office for National Statistics. *Coronavirus (COVID-19) related deaths by disability status, England and Wales: 24 January 2020 to 9 March 2022*. Office for National Statistics, 2022.
41. Matthew DB. *Just Medicine: A Cure for Racial Inequality in American Health Care*. NYU Press, 2015.
42. Institute of Medicine. *Unequal Treatment: Confronting Racial and Ethnic Disparities in Healthcare*. National Academies Press, 2003.
43. Cunningham TJ et al. Vital signs: racial disparities in age-specific mortality among blacks or African Americans—United States, 1999–2015. *MMWR Morb Mortal Wkly Rep* 2017; 66: 444.
44. Dwyer-Lindgren L et al. Ten Americas: a systematic analysis of life expectancy disparities in the USA. *The Lancet*, https://www.thelancet.com/journals/lancet/article/PIIS0140-6736(24)01495-8/fulltext (2024).
45. https://www.nytimes.com/interactive/2020/07/05/us/coronavirus-latinos-african-americans-cdc-data.html (Jul 5, 2020).
46. Klick J, Satel SL. *The Health Disparities Myth: Diagnosing the Treatment Gap*. AEI Press, 2006.
47. https://www.pewresearch.org/science/2022/04/07/black-americans-views-about-health-disparities-experiences-with-health-care/ (2022).
48. Gaskin DJ et al. Residential segregation and the availability of primary care physicians. *Health Serv Res* 2012; 47: 2353–2376.
49. Sager A. Why urban voluntary hospitals close. *Health Serv Res* 1983; 18: 451.
50. Eberth JM et al. The problem of the color line: spatial access to hospital services for minoritized racial and ethnic groups. *Health Aff (Millwood)* 2022; 41: 237–246.

3 Web-side Visits

1. Sund T, Rinde E. Telemedicine: still waiting for users. *The Lancet* 1995; 346: S24.
2. https://protomag.com/technology/doctoring-screen/ (2018).
3. Larkin M. Telemedicine links US doctors with space. *The Lancet* 1997; 349: 929.
4. https://abcnews.go.com/Technology/story?id=119358&page=1 (2000).
5. Field MJ, Grigsby J. Telemedicine and remote patient monitoring. *JAMA* 2002; 288: 423–425.
6. Mehrotra A et al. Rapidly converting to "virtual practices": outpatient care in the era of COVID-19. *NEJM Catal Innov Care Deliv*; 1.
7. Alexander GC et al. Use and content of primary care office-based vs telemedicine care visits during the COVID-19 pandemic in the US. *JAMA Netw Open* 2020; 3: e2021476.
8. Walley D et al. Use of telemedicine in general practice in Europe since the COVID-19 pandemic: a scoping review of patient and practitioner perspectives. *PLOS Digit Health* 2024; 3: e0000427.
9. https://www.bmj.com/content/371/bmj.m4348 (2020).
10. Kahn JM. Virtual visits—confronting the challenges of telemedicine. *N Engl J Med* 2015; 372: 1684–1685.
11. Ramaswamy A et al. Patient satisfaction with telemedicine during the COVID-19 pandemic: retrospective cohort study. *J Med Internet Res* 2020; 22: e20786.

12. https://hbr.org/2020/12/what-patients-like-and-dislike-about-telemedicine (2020).
13. https://media.npr.org/assets/img/2021/10/08/national-report-101221-final.pdf (2021).
14. Tieman JJ et al. Using telehealth to support end of life care in the community: a feasibility study. *BMC Palliat Care* 2016; 15: 1–7.
15. Layfield E et al. Telemedicine for head and neck ambulatory visits during COVID-19: evaluating usability and patient satisfaction. *Head Neck* 2020; 42: 1681–1689.
16. Triantafillou V et al. Patient perceptions of head and neck ambulatory telemedicine visits: a qualitative study. *Otolaryngol Neck Surg* 2021 May; 164(5): 923–931.
17. https://www.doctor.com/blog/telemedicine-today-patient-adoption (2020).
18. https://www.health.org.uk/news-and-comment/charts-and-infographics/how-does-uk-health-spending-compare-across-europe-over-the-past-decade (2022).
19. https://www.theguardian.com/society/2019/dec/23/uk-has-second-lowest-number-of-doctors-per-capita-in-europe (Dec 23, 2019).
20. https://www.cnbc.com/2020/04/09/telemedecine-demand-explodes-in-uk-as-gps-adapt-to-coronavirus-crisis.html (2020).
21. https://www.bmj.com/content/379/bmj.o2934 (2022).
22. Mroz G et al. UK newspapers "on the warpath": media analysis of general practice remote consulting in 2021. *Br J Gen Pract* 2022; 72: e907–e915.
23. https://www.capterra.co.uk/blog/2000/telemedicine-uk (2021).
24. https://www.fiercehealthcare.com/tech/demand-for-virtual-mental-health-soaring-here-are-notable-trends-who-using-it-and-why (2020).
25. https://www.cdc.gov/nchs/nvss/vsrr/drug-overdose-data.htm (2023).
26. Jones CM et al. Receipt of telehealth services, receipt and retention of medications for opioid use disorder, and medically treated overdose among Medicare beneficiaries before and during the COVID-19 pandemic. *JAMA Psychiatry* 2022 Oct 1; 79(10): 981–992.
27. Bashshur RL et al. The empirical evidence for telemedicine interventions in mental disorders. *Telemed E-Health* 2016; 22: 87–113.
28. Jenkins-Guarnieri MA et al. Patient perceptions of telemental health: systematic review of direct comparisons to in-person psychotherapeutic treatments. *Telemed E-Health* 2015; 21: 652–660.
29. Grubbs KM et al. A comparison of mental health diagnoses treated via interactive video and face to face in the Veterans Healthcare Administration. *Telemed E-Health* 2015; 21: 564–566.
30. Jacobs JC et al. Increasing mental health care access, continuity, and efficiency for veterans through telehealth with video tablets. *Psychiatr Serv* 2019; 70: 976–982.
31. https://oig.hhs.gov/reports/all/2022/telehealth-was-critical-for-providing-services-to-medicare-beneficiaries-during-the-first-year-of-the-covid-19-pandemic/ (2022).
32. Powell RE et al. Patient perceptions of telehealth primary care video visits. *Ann Fam Med* 2017; 15: 225–229.
33. Iezzoni LI, O'Day B. *More than Ramps: A Guide to Improving Health Care Quality and Access for People with Disabilities*. Oxford University Press, 2006.

34. https://www.ama-assn.org/practice-management/digital/telehealth-resource-center-research-findings (2023).
35. Dullet NW et al. Impact of a university-based outpatient telemedicine program on time savings, travel costs, and environmental pollutants. *Value Health* 2017; 20: 542–546.
36. Reed ME et al. Patient characteristics associated with choosing a telemedicine visit vs office visit with the same primary care clinicians. *JAMA Netw Open* 2020; 3: e205873.
37. Anastos-Wallen RE et al. Primary care appointment completion rates and telemedicine utilization among Black and non-Black patients from 2019 to 2020. *Telemed E-Health* 2022 Dec 1; 28(12): 1786–1795.
38. Jennett PA et al. The socio-economic impact of telehealth: a systematic review. *J Telemed Telecare* 2003; 9: 311–320.
39. Donelan K et al. Patient and clinician experiences with telehealth for patient follow-up care. *Am J Manag Care* 2019; 25: 40–44.
40. Zulman DM et al. Making connections: nationwide implementation of video telehealth tablets to address access barriers in veterans. *JAMIA Open* 2019; 2: 323–329.
41. https://eur-lex.europa.eu/legal-content/EN/ALL/?uri=celex%3A52008DC0689 (2008).
42. Cross M. Doctors are reluctant to use telemedicine and misunderstand what patients want, says NHS Confederation report. *BMJ Online* 2011; 342: d354.
43. https://www.bmj.com/content/367/bmj.l7087 (2019).
44. Fischer SH et al. Prevalence and characteristics of telehealth utilization in the United States. *JAMA Netw Open* 2020; 3: e2022302.
45. https://www.itu.int/en/ITU-D/Statistics/Documents/facts/FactsFigures2020.pdf (2020).
46. https://www.gsma.com/r/somic/ (2024).
47. https://www.weforum.org/stories/2024/09/2-5-billion-people-lack-internet-access-how-connectivity-can-unlock-their-potential/ (2024).
48. https://www.pewresearch.org/internet/fact-sheet/mobile/ (2021).
49. Valdez RS et al. Ensuring full participation of people with disabilities in an era of telehealth. *J Am Med Inform Assoc* 2021; 28: 389–392.
50. Kohli K et al. The digital divide in access to broadband internet and mental healthcare. *Nat Ment Health* 2024; 2: 88–95.
51. Sieck CJ et al. Digital inclusion as a social determinant of health. *NPJ Digit Med* 2021; 4: 1–3.
52. Rodriguez JA et al. Differences in the use of telephone and video telemedicine visits during the COVID-19 pandemic. *Am J Manag Care* 2021; 27: 21–26.
53. Ye S et al. Telemedicine expansion during the COVID-19 pandemic and the potential for technology-driven disparities. *J Gen Intern Med* 2021; 36: 256–258.
54. Raja M et al. Telehealth and digital developments in society that persons 75 years and older in European countries have been part of: a scoping review. *BMC Health Serv Res* 2021; 21: 1157.
55. Mao A et al. Barriers to telemedicine video visits for older adults in independent living facilities: mixed methods cross-sectional needs assessment. *JMIR Aging* 2022; 5: e34326.

56. Rodriguez JA et al. Digital inclusion as health care—supporting health care equity with digital-infrastructure initiatives. *N Engl J Med* 2022; 386: 1101–1103.
57. https://www.nejm.org/doi/full/10.1056/NEJMp2115646 (2022).
58. Cimperman M et al. Older adults' perceptions of home telehealth services. *Telemed E-Health* 2013; 19: 786–790.
59. https://commission.europa.eu/europes-digital-decade-digital-targets-2030-documents_en (2023).
60. Benjenk I et al. Disparities in audio-only telemedicine use among Medicare beneficiaries during the coronavirus disease 2019 pandemic. *Med Care* 2021; 59: 1014–1022.
61. https://www.bbc.co.uk/news/uk-england-leeds-64348478 (2023).
62. Payne R et al. Patient safety in remote primary care encounters: multimethod qualitative study combining Safety I and Safety II analysis. *BMJ Qual Saf* 2024; 33: 573–586.
63. Baughman DJ et al. Comparison of quality performance measures for patients receiving in-person vs telemedicine primary care in a large integrated health system. *JAMA Netw Open* 2022; 5: e2233267.
64. Demaerschalk BM et al. Assessment of clinician diagnostic concordance with video telemedicine in the integrated multispecialty practice at Mayo Clinic during the beginning of COVID-19 pandemic from March to June 2020. *JAMA Netw Open* 2022; 5: e2229958.
65. https://www.healthcareitnews.com/news/what-eventual-end-phe-would-mean-telehealth (2022).
66. https://www.who.int/europe/news-room/31-10-2022-telemedicine-has-clear-benefits-for-patients-in-european-countries--new-study-shows (2022).
67. Saigí-Rubió F et al. The current status of telemedicine technology use across the World Health Organization European region: an overview of systematic reviews. *J Med Internet Res* 2022; 24: e40877.

4 Doctor Deference

1. Goffman E. *The Presentation of Self in Everyday Life*. Penguin, 1978.
2. Malat JR et al. Race, socioeconomic status, and the perceived importance of positive self-presentation in health care. *Soc Sci Med* 2006; 62: 2479–2488.
3. Eysenck MW et al. Anxiety and cognitive performance: attentional control theory. *Emotion* 2007; 7: 336.
4. Jones W (ed). *Hippocrates II*. Harvard University Press, 1998, p. 297.
5. Solomon M. *Making Medical Knowledge*. Oxford University Press, 2015.
6. Castelo-Branco C et al. Do patients lie? An open interview vs. a blind questionnaire on sexuality. *J Sex Med* 2010; 7: 873–880.
7. Vogel L. Why do patients often lie to their doctors? *Can Med Assoc*, 2019; 191 (4): E115.
8. Pemberton M. *Trust Me, I'm a (Junior) Doctor*. Hachette UK, 2011.
9. Burgoon M et al. Patients who deceive: an empirical investigation of patient–physician communication. *J Lang Soc Psychol* 1994; 13: 443–468.
10. https://www.medicareadvantage.com/patient-doctor-lies-survey (2018).
11. Levy AG et al. Prevalence of and factors associated with patient nondisclosure of medically relevant information to clinicians. *JAMA Netw Open* 2018; 1: e185293.

12. Smith LK et al. Patients' help-seeking experiences and delay in cancer presentation: a qualitative synthesis. *The Lancet* 2005; 366: 825–831.
13. Whitaker KL et al. Help seeking for cancer "alarm" symptoms: a qualitative interview study of primary care patients in the UK. *Br J Gen Pr* 2015; 65: e96–e105.
14. Cromme SK et al. Worrying about wasting GP time as a barrier to help-seeking: a community-based, qualitative study. *Br J Gen Pr* 2016; 66: e474–e482.
15. Llanwarne N et al. Wasting the doctor's time? A video-elicitation interview study with patients in primary care. *Soc Sci Med* 2017; 176: 113–122.
16. Forbes LJL et al. Differences in cancer awareness and beliefs between Australia, Canada, Denmark, Norway, Sweden and the UK (the International Cancer Benchmarking Partnership): do they contribute to differences in cancer survival? *Br J Cancer* 2013; 108: 292.
17. Bell RA et al. Suffering in silence: reasons for not disclosing depression in primary care. *Ann Fam Med* 2011; 9: 439–446.
18. Levy AG et al. Assessment of patient nondisclosures to clinicians of experiencing imminent threats. *JAMA Netw Open* 2019; 2: e199277.
19. Freidson E. *Professional Dominance: The Social Structure of Medical Care*. Routledge, 2017.
20. Cheng JT et al. Two ways to the top: evidence that dominance and prestige are distinct yet viable avenues to social rank and influence. *J Pers Soc Psychol* 2013; 104: 103.
21. Barkow JH. Prestige and the ongoing process of culture revision. In: Cheng JT, Tracy J, Anderson C (eds) *The Psychology of Social Status*. Springer, 2014, pp. 29–45.
22. Henrich J, Gil-White FJ. The evolution of prestige: freely conferred deference as a mechanism for enhancing the benefits of cultural transmission. *Evol Hum Behav* 2001; 22: 165–196.
23. https://code-medical-ethics.ama-assn.org/ethics-opinions/gifts-patients (2022).
24. Fiske ST et al. A model of (often mixed) stereotype content: competence and warmth respectively follow from perceived status and competition. *J Pers Soc Psychol* 2002; 82: 878.
25. Arneill AB, Devlin AS. Perceived quality of care: the influence of the waiting room environment. *J Environ Psychol* 2002; 22: 345–360.
26. Devlin AS et al. "Impressive?" Credentials, family photographs, and the perception of therapist qualities. *J Environ Psychol* 2009; 29: 503–512.
27. https://www.bma.org.uk/news/2012/may/doctors-decide-what-not-to-wear (2012).
28. Petrilli CM et al. Understanding the role of physician attire on patient perceptions: a systematic review of the literature—targeting attire to improve likelihood of rapport (TAILOR) investigators. *BMJ Open* 2015; 5: e006578.
29. Houchens N et al. International patient preferences for physician attire: results from cross-sectional studies in four countries across three continents. *BMJ Open* 2022; 12: e061092.
30. Hall JA et al. Nonverbal behavior and the vertical dimension of social relations: a meta-analysis. *Psychol Bull* 2005; 131: 898.
31. Cheng JT et al. Listen, follow me: dynamic vocal signals of dominance predict emergent social rank in humans. *J Exp Psychol Gen* 2016; 145: 536.

32. Holland E et al. Visual attention to powerful postures: people avert their gaze from nonverbal dominance displays. *J Exp Soc Psychol* 2017; 68: 60–67.
33. Cheng JT, Tracy JL. The impact of wealth on prestige and dominance rank relationships. *Psychol Inq* 2013; 24: 102–108.
34. Lill MM, Wilkinson TJ. Judging a book by its cover: descriptive survey of patients' preferences for doctors' appearance and mode of address. *BMJ* 2005; 331: 1524–1527.
35. Ospina NS et al. Eliciting the patient's agenda—secondary analysis of recorded clinical encounters. *J Gen Intern Med* 2019; 34: 36–40.
36. Beckman HB, Frankel RM. The effect of physician behavior on the collection of data. *Ann Intern Med* 1984; 101: 692–696.
37. Marvel MK et al. Soliciting the patient's agenda: have we improved? *JAMA* 1999; 281: 283–287.
38. Mauksch LB. Questioning a taboo: physicians' interruptions during interactions with patients. *JAMA* 2017; 317: 1021–1022.
39. Langewitz W et al. Spontaneous talking time at start of consultation in outpatient clinic: cohort study. *BMJ* 2002; 325: 682–683.
40. Frankel RM. From sentence to sequence: understanding the medical encounter through microinteractional analysis. *Discourse Process* 1984; 7: 135–170.
41. Leape LL et al. Perspective: a culture of respect, Part 1 The nature and causes of disrespectful behavior by physicians. *Acad Med* 2012; 87: 845–852.
42. Leape LL et al. Perspective: a culture of respect, Part 2 Creating a culture of respect. *Acad Med* 2012; 87: 853–858.
43. Goldman B. Derogatory slang in the hospital setting. *AMA J Ethics* 2015; 17: 167–171.
44. Fox AT et al. Medical slang in British hospitals. *Ethics Behav* 2003; 13: 173–189.
45. Shem S. *The House of God*. Bantam Dell, 1978.
46. Goldman B. *The Secret Language of Doctors: Cracking the Code of Hospital Culture*. Triumph Books, 2015.
47. Park J et al. Physician use of stigmatizing language in patient medical records. *JAMA Netw Open* 2021; 4: e2117052.
48. Beach MC et al. Testimonial injustice: linguistic bias in the medical records of black patients and women. *J Gen Intern Med* 2021; 36: 1708–1714.
49. Blease C et al. Patients, clinicians and open notes: information blocking as a case of epistemic injustice. *J Med Ethics* 2022; 48: 785–793.
50. Scandurra I et al. Patient accessible EHR is controversial: lack of knowledge and diverse perceptions among professions. *Int J Reliab Qual E-Health IJRQEH* 2017; 6: 29–45.
51. https://projectdome.files.wordpress.com/2021/01/sustains-rapport.pdf (2014).
52. McCarthy DM et al. What did the doctor say? Health literacy and recall of medical instructions. *Med Care* 2012; 50: 277.
53. Kessels RP. Patients' memory for medical information. *J R Soc Med* 2003; 96: 219–222.
54. Blease C et al. Empowering patients and reducing inequities: is there potential in sharing clinical notes? *BMJ Qual Saf* 2020; 29(10): 1–2.

55. https://www.medtechpulse.com/article/infographic/americans-turning-to-the-internet-instead-of-a-physician (Oct 18, 2023).
56. https://ec.europa.eu/eurostat/web/products-eurostat-news/w/ddn-2022 1215-2 (2022).
57. https://www.usertesting.com/resources/library/industry-reports/us-consumer-perceptions-ai-healthcare (2024).
58. Silver MP. Patient perspectives on online health information and communication with doctors: a qualitative study of patients 50 years old and over. *J Med Internet Res* 2015; 17: e19.
59. Blease CR, Locher C, Gaab J et al. Generative artificial intelligence in primary care: an online survey of UK general practitioners. *BMJ Health Care Inform* 2024; 31(1): e101102.
60. Blease C, Worthen A, Torous J. Psychiatrists' experiences and opinions of generative artificial intelligence in mental healthcare: an online mixed methods survey. *Psychiatry Res* 2024; 333: 115724.
61. Blease C et al. Generative artificial intelligence in medicine: a mixed methods survey of UK general practitioners. Unpubl Manuscr, under review.
62. Tan SS-L, Goonawardene N. Internet health information seeking and the patient–physician relationship: a systematic review. *J Med Internet Res* 2017; 19: e9.
63. Stevenson FA et al. Information from the internet and the doctor–patient relationship: the patient perspective—a qualitative study. *BMC Fam Pract* 2007; 8: 47.
64. Broom A. Virtually healthy: the impact of internet use on disease experience and the doctor–patient relationship. *Qual Health Res* 2005; 15: 325–345.
65. Hart A et al. The role of the internet in patient–practitioner relationships: findings from a qualitative research study. *J Med Internet Res* 2004; 6: e36.
66. Chiu Y-C. Probing, impelling, but not offending doctors: the role of the internet as an information source for patients' interactions with doctors. *Qual Health Res* 2011; 21: 1658–1666.
67. Hay MC et al. Prepared patients: internet information seeking by new rheumatology patients. *Arthritis Care Res* 2008; 59: 575–582.
68. Clarke R. *Dear Life: A Doctor's Story of Love and Loss*. Little Brown, 2020.
69. Frosch DL et al. Authoritarian physicians and patients' fear of being labeled "difficult" among key obstacles to shared decision making. *Health Aff (Millwood)* 2012; 31: 1030–1038.
70. https://www.telegraph.co.uk/news/health/news/9568976/Patients-darent-complain-about-bad-GPs.html (Sept 27, 2012).
71. Doherty C, Stavropoulou C. Patients' willingness and ability to participate actively in the reduction of clinical errors: a systematic literature review. *Soc Sci Med* 2012; 75: 257–263.
72. https://www.cqc.org.uk/news/releases/new-research-cqc-shows-people-regret-not-raising-concerns-about-their-care-those-who (2019).
73. https://www.ombudsman.org.uk/publications/survey-experiences-nhs-mental-health care-england (2020).
74. Fisher KA et al. We want to know: patient comfort speaking up about breakdowns in care and patient experience. *BMJ Qual Saf* 2019; 28: 190–197.

5 Pouring Our Hearts Out to Machines

1. Slack WV. Patient counseling by computer. In: *Proceedings of the Annual Symposium on Computer Application in Medical Care*. American Medical Informatics Association, 1978, p. 222.
2. Slack WV et al. A computer-based medical-history system. *N Engl J Med* 1966; 274: 194–198.
3. Weizenbaum J. ELIZA—a computer program for the study of natural language communication between man and machine. *Commun ACM* 1966; 9: 36–45.
4. Overhage JM, McCallie Jr D. Physician time spent using the electronic health record during outpatient encounters: a descriptive study. *Ann Intern Med* 2020; 172: 169–174.
5. https://www.bmj.com/content/362/bmj.k3194 (2018).
6. Laranjo L et al. Conversational agents in healthcare: a systematic review. *J Am Med Inform Assoc* 2018; 25: 1248–1258.
7. Banks J et al. Use of an electronic consultation system in primary care: a qualitative interview study. *Br J Gen Pract* 2018; 68: e1–e8.
8. Tu T et al. Towards conversational diagnostic AI, http://arxiv.org/abs/2401.05654 (2024).
9. Bachman JW. The patient–computer interview: a neglected tool that can aid the clinician. *Mayo Clinic Proceedings*, 2003; 78(1): 67–78.
10. Zakim D et al. Underutilization of information and knowledge in everyday medical practice: evaluation of a computer-based solution. *BMC Med Inform Decis Mak* 2008; 8: 1–12.
11. Lucas RW et al. Psychiatrists and a computer as interrogators of patients with alcohol-related illnesses: a comparison. *Br J Psychiatry* 1977; 131: 160–167.
12. Skinner HA, Allen BA. Does the computer make a difference? Computerized versus face-to-face versus self-report assessment of alcohol, drug, and tobacco use. *J Consult Clin Psychol* 1983; 51: 267.
13. Locke SE et al. Computer-based interview for screening blood donors for risk of HIV transmission. *JAMA* 1992; 268: 1301–1305.
14. Greist JH et al. A computer interview for suicide-risk prediction. *Am J Psychiatry* 1973; 130: 1327–1332.
15. Slack WV. Patient–computer dialogue: a review. *Yearb Med Inform* 2000; 9: 71–78.
16. Richens J et al. A randomised controlled trial of computer-assisted interviewing in sexual health clinics. *Sex Transm Infect* 2010; 86: 310–314.
17. https://ncadv.org/STATISTICS (2022).
18. Hussain N et al. A comparison of the types of screening tool administration methods used for the detection of intimate partner violence: a systematic review and meta-analysis. *Trauma Violence Abuse* 2015; 16: 60–69.
19. Ahmad I et al. Intimate partner violence screening in emergency department: a rapid review of the literature. *J Clin Nurs* 2017; 26: 3271–3285.
20. Roter DL et al. Can e-mail messages between patients and physicians be patient-centered? *Health Commun* 2008; 23: 80–86.
21. Grønning A et al. How do patients and general practitioners in Denmark perceive the communicative advantages and disadvantages of access via email consultations? A media-theoretical qualitative study. *BMJ Open* 2020; 10: e039442.

22. Kindratt TB et al. Email patient–provider communication and cancer screenings among US adults: cross-sectional study. *JMIR Cancer* 2021; 7: e23790.
23. https://www.businesswire.com/news/home/20130730005132/en/Survey-Over-A-Third-of-Americans-Confess-to-Verbal-or-Physical-Abuse-of-Their-Computers (Jul 30, 2013).
24. Waytz A et al. Making sense by making sentient: effectance motivation increases anthropomorphism. *J Pers Soc Psychol* 2010; 99: 410.
25. Hume D. *The Natural History of Religion*. Stanford University Press, 1957.
26. https://www.dailystar.co.uk/news/latest-news/how-daily-stars-liz-vs-28289164 (Oct 20, 2022).
27. Gauthier I, Nelson CA. The development of face expertise. *Curr Opin Neurobiol* 2001; 11: 219–224.
28. Sperber D. *Explaining Culture: A Naturalistic Approach*. Blackwell, 1996.
29. Sperber D, Hirschfeld LA. The cognitive foundations of cultural stability and diversity. *Trends Cogn Sci* 2004; 8: 40–46.
30. Brown S. Where the wild brands are: some thoughts on anthropomorphic marketing. *Mark Rev* 2010; 10: 209–224.
31. Landwehr JR et al. It's got the look: the effect of friendly and aggressive "facial" expressions on product liking and sales. *J Mark* 2011; 75: 132–146.
32. Chandler J, Schwarz N. Use does not wear ragged the fabric of friendship: thinking of objects as alive makes people less willing to replace them. *J Consum Psychol* 2010; 20: 138–145.
33. Heider F, Simmel M. An experimental study of apparent behavior. *Am J Psychol* 1944; 57: 243–259.
34. https://www.youtube.com/watch?v=VTNmLt7QX8E.
35. Waytz A et al. Who sees human? The stability and importance of individual differences in anthropomorphism. *Perspect Psychol Sci* 2010; 5: 219–232.
36. Nowak KL, Rauh C. The influence of the avatar on online perceptions of anthropomorphism, androgyny, credibility, homophily, and attraction. *J Comput-Mediat Commun* 2005; 11: 153–178.
37. Koda T, Maes P. Agents with faces: the effect of personification. In: *Proceedings 5th IEEE International Workshop on Robot and Human Communication. RO-MAN'96 TSUKUBA*. IEEE, 1996, pp. 189–194.
38. Gazzola V et al. The anthropomorphic brain: the mirror neuron system responds to human and robotic actions. *Neuroimage* 2007; 35: 1674–1684.
39. Schuetzler RM et al. The impact of chatbot conversational skill on engagement and perceived humanness. *J Manag Inf Syst* 2020; 37: 875–900.
40. Sproull L et al. When the interface is a face. *Hum–Comput Interact* 1996; 11: 97–124.
41. Gratch J et al. It's only a computer: the impact of human–agent interaction in clinical interviews. In: *Proceedings of the 2014 International Conference on Autonomous Agents and Multi-agent Systems*. 2014, pp. 85–92.
42. Lucas GM et al. It's only a computer: virtual humans increase willingness to disclose. *Comput Hum Behav* 2014; 37: 94–100.
43. Walker JH et al. Using a human face in an interface. In: *Proceedings of the SIGCHI Conference on Human Factors in Computing Systems*. 1994, pp. 85–91.

44. Schuetzler RM et al. The influence of conversational agent embodiment and conversational relevance on socially desirable responding. *Decis Support Syst* 2018; 114: 94–102.
45. Schuetzler RM et al. The effect of conversational agent skill on user behavior during deception. *Comput Hum Behav* 2019; 97: 250–259.
46. Lalot F, Bertram A-M. When the bot walks the talk: investigating the foundations of trust in an artificial intelligence (AI) chatbot. *J Exp Psychol Gen*, https://psycnet.apa.org/record/2025-50901-001 (2025).
47. https://www.usertesting.com/resources/library/industry-reports/us-consumer-perceptions-ai-healthcare (2024).
48. Bornstein RF. Exposure and affect: overview and meta-analysis of research, 1968–1987. *Psychol Bull* 1989; 106: 265.
49. Montoya RM et al. A re-examination of the mere exposure effect: the influence of repeated exposure on recognition, familiarity, and liking. *Psychol Bull* 2017; 143: 459.
50. Mancia G et al. Cardiovascular risk associated with white-coat hypertension: pro side of the argument. *Hypertension* 2017; 70: 668–675.
51. https://www.commonwealthfund.org/international-health-policy-center/system-stats/annual-physician-visits-per-capita (2020).
52. Rachas A et al. Interactive telemedicine: effects on professional practice and health care outcomes. *Cochrane Database of Systematic Reviews* 2015(9).
53. Taylor ML et al. Does remote patient monitoring reduce acute care use? A systematic review. *BMJ Open* 2021; 11: e040232.
54. https://www.statista.com/statistics/1342873/daily-media-consumption-time-internet-users-wesern-europe/ (2025).
55. https://www.reviews.org/mobile/cell-phone-addiction/ (2025).
56. https://www.pewresearch.org/internet/2022/08/10/teens-social-media-and-technology-2022/ (2022).
57. Henson P et al. Impact of dynamic greenspace exposure on symptomatology in individuals with schizophrenia. *PLOS One* 2020; 15: e0238498.
58. Haidt J. *The Anxious Generation: How the Great Rewiring of Childhood Is Causing an Epidemic of Mental Illness*. Random House, 2024.
59. Blease CR. Too many "friends," too few "likes"? Evolutionary psychology and "Facebook depression." *Rev Gen Psychol* 2015; 19: 1–13.
60. Faurholt-Jepsen M et al. Voice analyses using smartphone-based data in patients with bipolar disorder, unaffected relatives and healthy control individuals, and during different affective states. *Int J Bipolar Disord* 2021; 9: 1–13.
61. Faurholt-Jepsen M et al. Objective smartphone data as a potential diagnostic marker of bipolar disorder. *Aust N Z J Psychiatry* 2019; 53: 119–128.
62. Pennebaker JW. The secret life of pronouns. *New Sci* 2011; 211: 42–45.
63. Eichstaedt JC et al. Facebook language predicts depression in medical records. *Proc Natl Acad Sci* 2018; 115: 11203–11208.
64. Agbavor F, Liang H. Predicting dementia from spontaneous speech using large language models. *PLOS Digital Health* 2022 Dec 22; 1(12): e0000168.
65. Reis BY et al. Longitudinal histories as predictors of future diagnoses of domestic abuse: modelling study. *BMJ* 2009 Sep 30; 339.
66. https://www.nimh.nih.gov/health/statistics/suicide (2022).

67. Luoma JB et al. Contact with mental health and primary care providers before suicide: a review of the evidence. *Am J Psychiatry* 2002; 159: 909–916.
68. Franklin JC et al. Risk factors for suicidal thoughts and behaviors: a meta-analysis of 50 years of research. *Psychol Bull* 2017; 143: 187.
69. Cusick M et al. Portability of natural language processing methods to detect suicidality from clinical text in US and UK electronic health records. *J Affect Disord Rep* 2022; 10: 100430.
70. Barak-Corren Y et al. Validation of an electronic health record-based suicide risk prediction modeling approach across multiple health care systems. *JAMA Netw Open* 2020; 3: e201262–e201262.
71. https://newatlas.com/health-wellbeing/computer-model-predict-suicide-attempts-machine-learning-health-records/ (2020).

6 Crafting Clinicians

1. Kahneman D, Sibony O, Sunstein CR. *Noise: A Flaw in Human Judgment*. Hachette UK, 2021. The present chapter elaborates on the extent, volume, and sources of medical noise, building on Kahneman, Sibony, and Sunstein's broader discussion.
2. McGlynn EA et al. The quality of health care delivered to adults in the United States. *N Engl J Med* 2003; 348: 2635–2645.
3. Levine DM et al. The quality of outpatient care delivered to adults in the United States, 2002 to 2013. *JAMA Intern Med* 2016; 176: 1778–1790.
4. Runciman WB et al. CareTrack: assessing the appropriateness of health care delivery in Australia. *Med J Aust* 2012; 197: 100–105.
5. Sheldon TA et al. What's the evidence that NICE guidance has been implemented? Results from a national evaluation using time series analysis, audit of patients' notes, and interviews. *BMJ* 2004; 329: 999.
6. Rosenkrantz AB et al. Discrepancy rates and clinical impact of imaging secondary interpretations: a systematic review and meta-analysis. *J Am Coll Radiol* 2018; 15: 1222–1231.
7. Johnson J, Kline JA. Intraobserver and interobserver agreement of the interpretation of pediatric chest radiographs. *Emerg Radiol* 2010; 17: 285–290.
8. Linder JA et al. Time of day and the decision to prescribe antibiotics. *JAMA Intern Med* 2014; 174: 2029–2031.
9. Hsiang EY et al. Association of primary care clinic appointment time with clinician ordering and patient completion of breast and colorectal cancer screening. *JAMA Netw Open* 2019; 2: e193403.
10. Ericsson A, Pool R. *Peak: Secrets from the New Science of Expertise*. Houghton Mifflin Harcourt, 2016.
11. Young JQ et al. "July effect": impact of the academic year-end changeover on patient outcomes: a systematic review. *Ann Intern Med* 2011; 155: 309–315.
12. Jena AB et al. Mortality among high-risk patients with acute myocardial infarction admitted to US teaching-intensive hospitals in July: a retrospective observational study. *Circulation* 2013; 128: 2754–2763.
13. Singh H et al. Medical errors involving trainees: a study of closed malpractice claims from 5 insurers. *Arch Intern Med* 2007; 167: 2030–2036.
14. Jeffe DB, Andriole DA. Factors associated with American Board of Medical Specialties member board certification among US medical school graduates. *JAMA* 2011; 306: 961–970.

15. Barnato AE et al. Hospital-level racial disparities in acute myocardial infarction treatment and outcomes. *Med Care* 2005; 43: 308.
16. Tsugawa Y et al. Comparison of hospital mortality and readmission rates for Medicare patients treated by male vs female physicians. *JAMA Intern Med* 2017; 177: 206–213.
17. Cucchetti A et al. The perceived ability of gastroenterologists, hepatologists and surgeons can bias medical decision making. *Int J Environ Res Public Health* 2020 Feb; 17(3): 1058.
18. Choudhry NK et al. Systematic review: the relationship between clinical experience and quality of health care. *Ann Intern Med* 2005; 142: 260–273.
19. Tsugawa Y et al. Physician age and outcomes in elderly patients in hospital in the US: observational study. *BMJ* 2017; 357: j1797.
20. Tsugawa Y et al. Age and sex of surgeons and mortality of older surgical patients: observational study. *BMJ* 2018 Apr 25; 361.
21. Eva KW. The aging physician: changes in cognitive processing and their impact on medical practice. *Acad Med* 2002; 77: S1–S6.
22. Dellinger EP et al. The aging physician and the medical profession: a review. *JAMA Surg* 2017; 152: 967–971.
23. Devi G. Alzheimer's disease in physicians—assessing professional competence and tempering stigma. *N Engl J Med* 2018; 378: 1073–1075.
24. Dijkstra AF et al. Gender bias in medical textbooks: examples from coronary heart disease, depression, alcohol abuse and pharmacology. *Med Educ* 2008; 42: 1021–1028.
25. https://www.bbc.com/news/av/uk-61258731 (May 1, 2022).
26. Cahill L. An issue whose time has come. *J Neurosci Res* 2017; 95: 12–13.
27. https://www.technologyreview.com/2022/08/15/1056908/biological-sex-immune-system/ (Aug 15, 2022).
28. Perez CC. *Invisible Women: Exposing Data Bias in a World Designed for Men*. Random House, 2019.
29. https://www.mirror.co.uk/3am/us-celebrity-news/lisa-marie-presley-ignored-several-30716292 (Aug 16, 2023).
30. https://www.mayoclinic.org/diseases-conditions/heart-disease/in-depth/heart-disease/art-20046167 (2019).
31. Bucholz EM et al. Editor's choice—sex differences in young patients with acute myocardial infarction: a VIRGO study analysis. *Eur Heart J Acute Cardiovasc Care* 2017; 6: 610–622.
32. Bontempo AC, Mikesell L. Patient perceptions of misdiagnosis of endometriosis: results from an online national survey. *Diagnosis* 2020; 7: 97–106.
33. https://committees.parliament.uk/publications/45909/documents/228040/default/ (Dec 11, 2024).
34. http://nationalpainreport.com/women-in-pain-report-significant-gender-bias-8824696.html (2014).
35. Schäfer G et al. Health care providers' judgments in chronic pain: the influence of gender and trustworthiness. *Pain* 2016; 157: 1618–1625.
36. Hirsh AT et al. The influence of patient's sex, race and depression on clinician pain treatment decisions. *Eur J Pain* 2013; 17: 1569–1579.
37. Epstein NK et al. Women's representation in RCTs evaluating FDA-supervised medical devices: a systematic review. *JAMA Intern Med* 2024; 184(8): 977–979.

38. Tharpe N. Adverse drug reactions in women's health care. *J Midwifery Women's Health* 2011; 56: 205–213.
39. Zucker I, Prendergast BJ. Sex differences in pharmacokinetics predict adverse drug reactions in women. *Biol Sex Differ* 2020; 11: 1–14.
40. Strother E et al. Eating disorders in men: underdiagnosed, undertreated, and misunderstood. *Eat Disord* 2012; 20: 346–355.
41. Wang F et al. Overall mortality after diagnosis of breast cancer in men vs women. *JAMA Oncol* 2019; 5: 1589–1596.
42. Keller, I. et al. *Global Survey on Geriatrics in the Medical Curriculum*. World Health Organization, 2002.
43. Mateos-Nozal J et al. A systematic review of surveys on undergraduate teaching of Geriatrics in medical schools in the XXI century. *Eur Geriatr Med* 2014; 5: 119–124.
44. https://kffhealthnews.org/news/article/who-will-care-for-older-adults-weve-plenty-of-know-how-but-too-few-specialists/ (2023).
45. Wyman MF et al. Ageism in the health care system: providers, patients, and systems. In: Ayalon L, Tesch-Römer C (eds) *Contemporary Perspectives on Ageism*. Springer, 2018, pp. 193–212.
46. Dunn C et al. Older cancer patients in cancer clinical trials are underrepresented. Systematic literature review of almost 5000 meta- and pooled analyses of phase III randomized trials of survival from breast, prostate and lung cancer. *Cancer Epidemiol* 2017; 51: 113–117.
47. Bourgeois FT et al. Prevalence and characteristics of interventional trials conducted exclusively in elderly persons: a cross-sectional analysis of registered clinical trials. *PLOS One* 2016 May 19; 11(5): e0155948.
48. Watts G. Why the exclusion of older people from clinical research must stop. *BMJ* 2012; 344: e3445.
49. Fialová D et al. Ageism in medication use in older patients. In: Ayalon L, Tesch-Römer C (eds) *Contemporary Perspectives on Ageism*. Springer, 2018, pp. 213–240.
50. Bonham VL et al. Examining how race, ethnicity, and ancestry data are used in biomedical research. *JAMA* 2018; 320: 1533–1534.
51. https://nap.nationalacademies.org/catalog/26479/improving-representation-in-clinical-trials-and-research-building-research-equity (2022).
52. Gottlieb ER et al. Assessment of racial and ethnic differences in oxygen supplementation among patients in the intensive care unit. *JAMA Intern Med* 2022; 182: 849–858.
53. Peterson PN et al. A validated risk score for in-hospital mortality in patients with heart failure from the American Heart Association Get With the Guidelines program. *Circ Cardiovasc Qual Outcomes* 2010; 3: 25–32.
54. Eberly LA et al. Identification of racial inequities in access to specialized inpatient heart failure care at an academic medical center. *Circ Heart Fail* 2019; 12: e006214.
55. Hoffman KM et al. Racial bias in pain assessment and treatment recommendations, and false beliefs about biological differences between blacks and whites. *Proc Natl Acad Sci* 2016; 113: 4296–4301.
56. Haendel M et al. *How Many Rare Diseases Are There?* Nature Publishing Group, 2019.
57. https://www.mda.org/disease/myotonic-dystrophy (2024).
58. https://www.globalgenes.org/rare-disease-facts/ (2024).

59. Commission of the European Communities. Commission notice on the application of Articles 3, 5 and 7 of Regulation (EC) No 141/2000 on orphan medicinal products. *Off J Eur Union* 2016 Nov 18; C 424: 3–9.
60. Richter T et al. Rare disease terminology and definitions—a systematic global review: report of the ISPOR rare disease special interest group. *Value Health* 2015; 18: 906–914.
61. Molster C et al. Survey of healthcare experiences of Australian adults living with rare diseases. *Orphanet J Rare Dis* 2016; 11: 30.
62. https://rarediseases.org/wp-content/uploads/2020/11/NRD-2088-Barriers-30-Yr-Survey-Report_FNL-2.pdf (2020).
63. Vickers PJ. Challenges and opportunities in the treatment of rare diseases. *Drug Discov World* 2013; 14: 9–16.
64. https://thezebranetwork.org/ (2024).
65. Rees CA et al. Noncompletion and nonpublication of trials studying rare diseases: a cross-sectional analysis. *PLOS Medicine* 2019 Nov 21; 16(11): e1002966.
66. Smith R. Thoughts for new medical students at a new medical school. *BMJ* 2003; 327: 1430–1433.
67. Ioannidis JP. Contradicted and initially stronger effects in highly cited clinical research. *JAMA* 2005; 294: 218–228.
68. Prasad VK, Cifu AS. *Ending Medical Reversal: Improving Outcomes, Saving Lives.* JHU Press, 2015.
69. Morris ZS et al. The answer is 17 years, what is the question: understanding time lags in translational research. *J R Soc Med* 2011; 104: 510–520.
70. Saint S et al. Journal reading habits of internists. *J Gen Intern Med* 2000; 15: 881–884.
71. Burke DT et al. Reading habits of practicing physiatrists. *Am J Phys Med Rehabil* 2002; 81: 779–787.
72. Schiff GD. Minimizing diagnostic error: the importance of follow-up and feedback. *Am J Med* 2008; 121: S38–S42.
73. Bloom BS. Effects of continuing medical education on improving physician clinical care and patient health: a review of systematic reviews. *Int J Technol Assess Health Care* 2005; 21: 380–385.
74. Forsetlund L et al. Continuing education meetings and workshops: effects on professional practice and healthcare outcomes. *Cochrane Database of Systematic Reviews* 2021(9).

7 The Dark Art of Medicine

1. https://www.vogue.com/article/serena-williams-vogue-cover-interview-february-2018 (Jan 10, 2018).
2. Wirth K, Scheibenbogen C. A unifying hypothesis of the pathophysiology of myalgic encephalomyelitis/chronic fatigue syndrome (ME/CFS): recognitions from the finding of autoantibodies against ß2-adrenergic receptors. *Autoimmun Rev* 2020; 19(6): 102527.
3. Klick J, Satel SL. *The Health Disparities Myth: Diagnosing the Treatment Gap.* AEI Press, 2006.
4. Olah ME et al. The effect of socioeconomic status on access to primary care: an audit study. *CMAJ* 2013; 185: e263–e269.

5. Kugelmass H. "Sorry, I'm not accepting new patients": an audit study of access to mental health care. *J Health Soc Behav* 2016; 57: 168–183.
6. Woo JK et al. Effect of patient socioeconomic status on perceptions of first- and second-year medical students. *CMAJ* 2004; 170: 1915–1919.
7. Arpey NC et al. How socioeconomic status affects patient perceptions of health care: a qualitative study. *J Prim Care Community Health* 2017; 8: 169–175.
8. Hausmann LR et al. Impact of perceived discrimination in health care on patient–provider communication. *Med Care* 2011; 49: 626.
9. Meyers DS et al. Primary care physicians' perceptions of the effect of insurance status on clinical decision making. *Ann Fam Med* 2006; 4: 399–402.
10. Robb K et al. Public awareness of cancer in Britain: a population-based survey of adults. *Nat Preced* 2009; 101 Suppl 2: S18–23.
11. Siminoff LA et al. Cancer communication patterns and the influence of patient characteristics: disparities in information-giving and affective behaviors. *Patient Educ Couns* 2006; 62: 355–360.
12. Mehra R et al. Racial residential segregation and adverse birth outcomes: a systematic review and meta-analysis. *Soc Sci Med* 2017; 191: 237–250.
13. Gujral K et al. Severe maternal morbidity and potential disparities by race and rurality among VA-covered births, 2010 to 2020. *Am J Obstet Gynecol*, https://doi.org/10.1016/j.ajog.2024.12.021 (2025).
14. Van Daalen KR et al. Racial discrimination and adverse pregnancy outcomes: a systematic review and meta-analysis. *BMJ Glob Health* 2022; 7: e009227.
15. https://www.pewresearch.org/social-trends/2019/04/09/race-in-america-2019/psdt_04-09-19_race-00-03/ (2019).
16. https://www.pewresearch.org/science/2020/05/21/trust-in-medical-scientists-has-grown-in-u-s-but-mainly-among-democrats/ (2020).
17. Sun M et al. Negative patient descriptors: documenting racial bias in the electronic health record. *Health Aff (Millwood)* 2022; 41: 203–211.
18. Pinder RJ et al. Minority ethnicity patient satisfaction and experience: results of the National Cancer Patient Experience Survey in England. *BMJ Open* 2016; 6: e011938.
19. Vedam S et al. The Giving Voice to Mothers study: inequity and mistreatment during pregnancy and childbirth in the United States. *Reprod Health* 2019; 16: 1–18.
20. https://www.birthrights.org.uk/wp-content/uploads/2022/05/Birthrights-inquiry-systemic-racism_exec-summary_May-22-web.pdf (May 2022).
21. FitzGerald C, Hurst S. Implicit bias in healthcare professionals: a systematic review. *BMC Med Ethics* 2017; 18: 19.
22. Landon BE et al. Assessment of racial disparities in primary care physician specialty referrals. *JAMA Netw Open* 2021; 4: e2029238.
23. Institute of Medicine. *Unequal Treatment: Confronting Racial and Ethnic Disparities in Healthcare*. National Academies Press, 2004.
24. Parikh K et al. Disparities in racial, ethnic, and payer groups for pediatric safety events in US hospitals. *Pediatrics* 2024 Mar 1; 153(3): e2023063714.
25. Singhal A et al. Racial-ethnic disparities in opioid prescriptions at emergency department visits for conditions commonly associated with prescription drug abuse. *PLOS One* 2016; 11: e0159224.

26. Hassan AM et al. Association between patient–surgeon race and gender concordance and patient-reported outcomes following breast cancer surgery. *Breast Cancer Res Treat* 2023; 198(1): 167–175.
27. Jetty A et al. Patient–physician racial concordance associated with improved healthcare use and lower healthcare expenditures in minority populations. *J Racial Ethn Health Disparities* 2022; 9: 68–81.
28. Alsan M et al. Does diversity matter for health? Experimental evidence from Oakland. *Am Econ Rev* 2019; 109: 4071–4111.
29. Shen MJ et al. The effects of race and racial concordance on patient–physician communication: a systematic review of the literature. *J Racial Ethn Health Disparities* 2018; 5: 117–140.
30. https://donoharmmedicine.org/wp-content/uploads/2023/12/Racial-Concordance-in-Medicine-The-Return-of-Segregation.pdf (2023).
31. Gordon HS et al. Racial differences in doctors' information-giving and patients' participation. *Cancer* 2006; 107: 1313–1320.
32. Street Jr RL et al. Physicians' communication and perceptions of patients: is it how they look, how they talk, or is it just the doctor? *Soc Sci Med* 2007; 65: 586–598.
33. Album D et al. Stability and change in disease prestige: a comparative analysis of three surveys spanning a quarter of a century. *Soc Sci Med* 2017; 180: 45–51.
34. Album D, Westin S. Do diseases have a prestige hierarchy? A survey among physicians and medical students. *Soc Sci Med* 2008; 66: 182–188.
35. Blease C et al. Epistemic injustice in healthcare encounters: evidence from chronic fatigue syndrome. *J Med Ethics* 2017; 43: 549–557.
36. https://www.meaction.net/wp-content/uploads/2019/10/Your-experience-of-ME-services-Survey-report-by-MEAction-UK.pdf (Oct, 2019).
37. Raine R et al. General practitioners' perceptions of chronic fatigue syndrome and beliefs about its management, compared with irritable bowel syndrome: qualitative study. *BMJ* 2004; 328: 1354–1357.
38. Ucok A. Other people stigmatize ... but, what about us? Attitudes of mental health professionals towards patients with schizophrenia. *Arch Neuropsychiatry* 2007; 44: 108–116.
39. Thornicroft G et al. Discrimination in health care against people with mental illness. *Int Rev Psychiatry* 2007; 19: 113–122.
40. Shefer G et al. Diagnostic overshadowing and other challenges involved in the diagnostic process of patients with mental illness who present in emergency departments with physical symptoms—a qualitative study. *PLOS One* 2014; 9: e111682.
41. Sullivan G et al. Disparities in hospitalization for diabetes among persons with and without co-occurring mental disorders. *Psychiatr Serv* 2006; 57: 1126–1131.
42. Himmelstein G et al. Examination of stigmatizing language in the electronic health record. *JAMA Netw Open* 2022; 5: e2144967.
43. Glassberg J et al. Among emergency physicians, use of the term "sickler" is associated with negative attitudes toward people with sickle cell disease. *Am J Hematol* 2013; 88: 532–533.
44. Iezzoni LI et al. Treatment disparities for disabled Medicare beneficiaries with stage I non-small cell lung cancer. *Arch Phys Med Rehabil* 2008; 89: 595–601.

45. Iezzoni LI. Eliminating health and health care disparities among the growing population of people with disabilities. *Health Aff (Millwood)* 2011; 30: 1947–1954.
46. McCarthy EP et al. Disparities in breast cancer treatment and survival for women with disabilities. *Ann Intern Med* 2006; 145: 637–645.
47. https://covid-drm.org/en/statements/covid-19-disability-rights-monitor-report-highlights-catastrophic-global-failure-to-protect-the-rights-of-persons-with-disabilities (2020).
48. Emanuel EJ et al. Fair allocation of scarce medical resources in the time of Covid-19. *NEJM* 2020 May 21; 382(21): 2049–2055.
49. Haque OS, Stein MA. COVID-19 clinical bias, persons with disabilities, and human rights. *Health Hum Rights* 2020; 22: 285–290.
50. Iezzoni LI et al. Mobility impairments and use of screening and preventive services. *Am J Public Health* 2000; 90: 955–961.
51. Iezzoni LI et al. Physicians' perceptions of people with disability and their health care. *Health Aff (Millwood)* 2021 Feb; 40(2): 297–306.
52. Agaronnik N et al. Exploring issues relating to disability cultural competence among practicing physicians. *Disabil Health J* 2019; 12: 403–410.
53. Bernstein C. Age and race fears seen in housing opposition. *The Washington Post*, Mar 7, 1969.
54. Greene MG et al. Ageism in the medical encounter: an exploratory study of the doctor–elderly patient relationship. *Lang Commun* 1986 Jan 1; 6(1–2): 113–124.
55. Lagacé M et al. The silent impact of ageist communication in long term care facilities: elders' perspectives on quality of life and coping strategies. *J Aging Stud* 2012; 26: 335–342.
56. Ben-Harush A et al. Ageism among physicians, nurses, and social workers: findings from a qualitative study. *Eur J Ageing* 2017; 14: 39–48.
57. Gould ON et al. Recall and subjective reactions to speaking styles: does age matter? *Exp Aging Res* 2002; 28: 199–213.
58. Ryan EB et al. Patronizing the old: how do younger and older adults respond to baby talk in the nursing home? *Int J Aging Hum Dev* 1994; 39: 21–32.
59. Peake MD et al. Ageism in the management of lung cancer. *Age Ageing* 2003; 32: 171–177.
60. Bhalla A et al. Older stroke patients in Europe: stroke care and determinants of outcome. *Age Ageing* 2004; 33: 618–624.
61. Jackson SE et al. Perceived weight discrimination and changes in weight, waist circumference, and weight status. *Obesity* 2014; 22: 2485–2488.
62. Ludwig DS et al. The carbohydrate-insulin model: a physiological perspective on the obesity pandemic. *Am J Clin Nutr* 2021; 114: 1873–1885.
63. Cooper CB et al. Sleep deprivation and obesity in adults: a brief narrative review. *BMJ Open Sport Exerc Med* 2018; 4: e000392.
64. O'Rahilly S, Farooqi IS. Human obesity: a heritable neurobehavioral disorder that is highly sensitive to environmental conditions. *Diabetes* 2008; 57: 2905–2910.
65. Van Leeuwen F et al. Is obesity stigma based on perceptions of appearance or character? Theory, evidence, and directions for further study. *Evol Psychol* 2015; 13(3): 1–11.

66. Bocquier A et al. Overweight and obesity: knowledge, attitudes, and practices of general practitioners in France. *Obes Res* 2005; 13: 787–795.
67. Fogelman Y et al. Managing obesity: a survey of attitudes and practices among Israeli primary care physicians. *Int J Obes* 2002; 26: 1393.
68. Hebl MR, Xu J. Weighing the care: physicians' reactions to the size of a patient. *Int J Obes* 2001; 25: 1246.
69. Foster GD et al. Primary care physicians' attitudes about obesity and its treatment. *Obes Res* 2003; 11: 1168–1177.
70. Phelan SM et al. Implicit and explicit weight bias in a national sample of 4,732 medical students: the medical student CHANGES study. *Obesity* 2014; 22: 1201–1208.
71. Brown I et al. Primary care support for tackling obesity: a qualitative study of the perceptions of obese patients. *Br J Gen Pr* 2006; 56: 666–672.
72. Bertakis KD, Azari R. The impact of obesity on primary care visits. *Obes Res* 2005; 13: 1615–1623.
73. Takeshita J et al. Association of racial/ethnic and gender concordance between patients and physicians with patient experience ratings. *JAMA Netw Open* 2020; 3: e2024583.
74. Hall JA et al. Patients' satisfaction with male versus female physicians: a meta-analysis. *Med Care* 2011; 49: 611–617.
75. Sandhu H et al. The impact of gender dyads on doctor–patient communication: a systematic review. *Patient Educ Couns* 2009; 76: 348–355.
76. Greenwood BN et al. Patient–physician gender concordance and increased mortality among female heart attack patients. *Proc Natl Acad Sci* 2018; 115: 8569–8574.
77. Thompson CM et al. Women's experiences of health-related communicative disenfranchisement. *Health Commun* 2022; 38: 3135–3146.
78. https://physicians.dukehealth.org/articles/recognizing-addressing-unintended-gender-bias-patient-care (2020).
79. Beach MC et al. Testimonial injustice: linguistic bias in the medical records of black patients and women. *J Gen Intern Med* 2021; 36: 1708–1714.
80. Charlesworth TE et al. Patterns of implicit and explicit attitudes: I. Long-term change and stability from 2007 to 2016. *Psychol Sci* 2019; 30: 174–192.
81. https://www.americanprogress.org/issues/lgbtq-rights/news/2018/01/18/445130/discrimination-prevents-lgbtq-people-accessing-health-care/ (2018).
82. Lippens L et al. The state of hiring discrimination: a meta-analysis of (almost) all recent correspondence experiments. *Eur Econ Rev* 2023; 151: 104315.
83. Kinzler KD. *How You Say It: Why You Talk the Way You Do—And What It Says About You*. Houghton Mifflin Harcourt, 2020.
84. https://www.youtube.com/watch?v=mbFFI18q9Zs&t=39s (2007).
85. https://www.theguardian.com/music/2019/oct/31/a-duel-with-van-morrison-is-this-a-psychiatric-examination-it-sounds-like-one (Oct 31, 2019).
86. https://www.youtube.com/watch?v=NCPdeowvA_M (2011).
87. Lakoff G. *Women, Fire, and Dangerous Things: What Categories Reveal about the Mind*. University of Chicago Press, 2008.
88. Kostopoulou O et al. Information distortion in physicians' diagnostic judgments. *Med Decis Making* 2012; 32: 831–839.

89. Pinker S. *How the Mind Works*. W.W. Norton & Co., 1997.
90. https://naacp.org/resources/criminal-justice-fact-sheet (2025).
91. https://www.migrationpolicy.org/article/sub-saharan-african-immigrants-united-states-2019 (May 11, 2022).
92. https://osr.statisticsauthority.gov.uk/wp-content/uploads/2021/09/Review-of-mental-health-statistics-in-Northern-Ireland.pdf (2021).
93. https://spsp.org/news-center/character-context-blog/stereotype-accuracy-one-largest-and-most-replicable-effects-all (2016).
94. Liu A, Xie Y. Why do Asian Americans academically outperform Whites? The cultural explanation revisited. *Soc Sci Res* 2016; 58: 210–226.
95. Kessler RC. Epidemiology of women and depression. *J Affect Disord* 2003; 74: 5–13.
96. Xie Z et al. Racial and ethnic disparities in medication adherence among privately insured patients in the United States. *PLOS One* 2019; 14: e0212117.
97. Semahegn A et al. Psychotropic medication non-adherence and its associated factors among patients with major psychiatric disorders: a systematic review and meta-analysis. *Syst Rev* 2020; 9: 1–18.
98. Park JH, Schaller M. Does attitude similarity serve as a heuristic cue for kinship? Evidence of an implicit cognitive association. *Evol Hum Behav* 2005; 26: 158–170.
99. Pelham BW et al. Why Susie sells seashells by the seashore: implicit egotism and major life decisions. *J Pers Soc Psychol* 2002; 82: 469.
100. Hewstone M et al. Intergroup bias. *Annu Rev Psychol* 2002; 53: 575–604.
101. Rivera LA. Hiring as cultural matching: the case of elite professional service firms. *Am Sociol Rev* 2012; 77: 999–1022.
102. Barkow JH. Introduction: sometimes the bus does wait. *Missing Revolut Darwinism Soc Sci* 2006; 3–60.
103. Pinker S. *The Better Angels of Our Nature: Why Violence has Declined*. Penguin Group USA, 2012.
104. Kurzban R et al. Can race be erased? Coalitional computation and social categorization. *Proc Natl Acad Sci* 2001; 98: 15387–15392.
105. Voorspoels W et al. Can race really be erased? A pre-registered replication study. *Front Psychol* 2014; 5: 1035.
106. Kahneman D. *Thinking, Fast and Slow*. Macmillan, 2011.
107. McCauley RN. *Why Religion Is Natural and Science Is Not*. Oxford University Press, 2011.
108. https://code-medical-ethics.ama-assn.org/ethics-opinions/disparities-health-care (2016).
109. FitzGerald C et al. Interventions designed to reduce implicit prejudices and implicit stereotypes in real world contexts: a systematic review. *BMC Psychol* 2019; 7: 29.
110. https://musaalgharbi.com/2020/09/16/diversity-important-related-training-terrible/ (2020).
111. Duguid MM, Thomas-Hunt MC. Condoning stereotyping? How awareness of stereotyping prevalence impacts expression of stereotypes. *J Appl Psychol* 2015; 100: 343.

112. Payne BK et al. Best laid plans: effects of goals on accessibility bias and cognitive control in race-based misperceptions of weapons. *J Exp Soc Psychol* 2002; 38: 384–396.
113. Legault L et al. Ironic effects of antiprejudice messages: how motivational interventions can reduce (but also increase) prejudice. *Psychol Sci* 2011; 22: 1472–1477.

8 Building Digital Doctors

1. Kung TH et al. Performance of ChatGPT on USMLE: potential for AI-assisted medical education using large language models. *PLOS Digit Health* 2023; 2: e0000198.
2. Haze T et al. Influence on the accuracy in ChatGPT: differences in the amount of information per medical field. *Int J Med Inf* 2023; 180: 105283.
3. Kanjee Z et al. Accuracy of a generative artificial intelligence model in a complex diagnostic challenge. *JAMA* 2023 Jul 3; 330: 78–80.
4. Eriksen AV et al. Use of GPT-4 to diagnose complex clinical cases. *NEJM AI*; 1. Epub ahead of print Jan 2024. doi: 10.1056/AIp2300031.
5. https://the-decoder.com/gpt-4-has-a-trillion-parameters/ (Mar 25, 2023).
6. Singhal K et al. Large language models encode clinical knowledge. *Nature* 2023; 620: 172–180.
7. Goffman E. *The Presentation of Self in Everyday Life*. Penguin, 1978.
8. Al-Dujaili Z et al. Assessing the accuracy and consistency of ChatGPT in clinical pharmacy management: a preliminary analysis with clinical pharmacy experts worldwide. *Res Soc Adm Pharm* 2023; 19: 1590–1594.
9. Morris ZS et al. The answer is 17 years, what is the question: understanding time lags in translational research. *J R Soc Med* 2011; 104: 510–520.
10. Ozgor BY, Simavi MA. Accuracy and reproducibility of ChatGPT's free version answers about endometriosis. *Int J Gynecol Obstet* 2023; ijgo.15309.
11. Mello MM, Guha N. ChatGPT and physicians' malpractice risk. *JAMA Health Forum* 2023 May 5; 4(5): e231938.
12. https://www.today.com/health/mom-chatgpt-diagnosis-pain-rcna101843 (2023).
13. Mehnen L et al. ChatGPT as a medical doctor? A diagnostic accuracy study on common and rare diseases. *medRxiv* 2023; doi:10.1101/2023.04.20.23288859.
14. Alsentzer E et al. Deep learning for diagnosing patients with rare genetic diseases. *medRxiv* 2022; doi: 10.1101/2022.12.07.22283238.
15. Yagin FH et al. An explainable artificial intelligence model proposed for the prediction of myalgic encephalomyelitis/chronic fatigue syndrome and the identification of distinctive metabolites. *Diagnostics* 2023; 13: 3495.
16. Jin Q et al. Matching patients to clinical trials with large language models. *Nat Commun* 2024; 15: 9074.
17. Wojtara M et al. Artificial intelligence in rare disease diagnosis and treatment. *Clin Transl Sci* 2023; 16: 2106–2111.
18. https://matt.might.net/articles/my-sons-killer/ (2012).
19. O'Neil C. *Weapons of Math Destruction: How Big Data Increases Inequality and Threatens Democracy*. Broadway Books, 2016.
20. https://www.gov.uk/government/groups/equity-in-medical-devices-independent-review (Mar 11, 2024).

21. Celi LA et al. Sources of bias in artificial intelligence that perpetuate healthcare disparities—a global review. *PLOS Digit Health* 2022; 1: e0000022.
22. Samaan JS et al. ChatGPT's ability to comprehend and answer cirrhosis related questions in Arabic. *Arab J Gastroenterol* 2023; 24: 145–148.
23. Menezes MCS et al. The potential of Generative Pre-trained Transformer 4 (GPT-4) to analyse medical notes in three different languages: a retrospective model-evaluation study. *Lancet Digit Health* 2025; 7: e35–e43.
24. Obermeyer Z et al. Dissecting racial bias in an algorithm used to manage the health of populations. *Science* 2019; 366: 447–453.
25. Omiye JA et al. Large language models propagate race-based medicine. *NPJ Digit Med* 2023; 6: 195.
26. Zack T et al. Assessing the potential of GPT-4 to perpetuate racial and gender biases in health care: a model evaluation study. *Lancet Digit Health* 2024; 6: e12–e22.
27. Movva R et al. Topics, authors, and networks in large language model research: trends from a survey of 17K arXiv papers. http://arxiv.org/abs/2307.10700 (2023).
28. Broussard M. *More than a Glitch: Confronting Race, Gender, and Ability Bias in Tech.* MIT Press, 2023.
29. Pierson E et al. Use large language models to promote equity, http://arxiv.org/abs/2312.14804 (2023).
30. Ferryman K et al. Considering biased data as informative artifacts in AI-assisted health care. *N Engl J Med* 2023; 389: 833–838.
31. Zou J et al. Implications of predicting race variables from medical images. *Science* 2023; 381: 149–150.
32. Buckley T et al. Accuracy of a vision-language model on challenging medical cases, http://arxiv.org/abs/2311.05591 (2023).
33. Himmelstein G et al. Examination of stigmatizing language in the electronic health record. *JAMA Netw Open* 2022; 5: e2144967.
34. Sun M et al. Negative patient descriptors: documenting racial bias in the electronic health record. *Health Aff (Millwood)* 2022 Feb; 41: 203–211.
35. Vyas DA et al. Hidden in plain sight—reconsidering the use of race correction in clinical algorithms. *N Engl J Med* 2020; 383: 874–882.
36. Pierson E. Accuracy and equity in clinical risk prediction. *N Engl J Med* 2024; 390: 100–102.
37. Buckley A et al. Racial and ethnic disparities among women undergoing a trial of labor after cesarean delivery: performance of the VBAC calculator with and without patients' race/ethnicity. *Reprod Sci* 2022; 29: 2030–2038.
38. Khor S et al. Racial and ethnic bias in risk prediction models for colorectal cancer recurrence when race and ethnicity are omitted as predictors. *JAMA Netw Open* 2023; 6: e2318495.
39. Zink A et al. Race corrections in clinical models: examining family history and cancer risk. *medRxiv* 2023; doi: 10.1101/2023.03.31.23287926.
40. Diao JA et al. Implications of race adjustment in lung-function equations. *N Engl J Med*, 2024 Jun 13; 390: 2083–2097.
41. Gichoya JW et al. AI recognition of patient race in medical imaging: a modelling study. *Lancet Digit Health* 2022; 4: e406–e414.
42. Yang Y et al. The limits of fair medical imaging AI in the wild. *arXiv* preprint arXiv:2312.10083. 2023 Dec 11.

43. Pierson E et al. An algorithmic approach to reducing unexplained pain disparities in underserved populations. *Nat Med* 2021; 27: 136–140.
44. https://www.medpagetoday.com/opinion/faustfiles/102723 (2023).
45. https://www.statnews.com/2023/02/01/promises-pitfalls-chatgpt-assisted-medicine/ (2023).
46. Chen A, Chen DO. Accuracy of chatbots in citing journal articles. *JAMA Netw Open* 2023; 6: e2327647.
47. Goddard J. Hallucinations in ChatGPT: a cautionary tale for biomedical researchers. *Am J Med*, 2023 Nov 1; 136(11): 1059–1060.
48. Tang L et al. Evaluating large language models on medical evidence summarization. *Npj Digit Med* 2023; 6: 158.
49. Bender EM et al. On the dangers of stochastic parrots: can language models be too big? In: *Proceedings of the 2021 ACM Conference on Fairness, Accountability, and Transparency*. Virtual Event Canada: ACM, pp. 610–623.
50. Chen S et al. Use of artificial intelligence chatbots for cancer treatment information. *JAMA Oncol* 2023 Oct 1; 9(10): 1459–1462.
51. Woo KC et al. Evaluation of GPT-4 ability to identify and generate patient instructions for actionable incidental radiology findings. *J Am Med Inform Assoc* 2024 Sept; 31: 1983–1993.
52. https://www.tomshardware.com/tech-industry/artificial-intelligence/concerns-about-medical-note-taking-tool-raised-after-researcher-discovers-it-invents-things-no-one-said-nabla-is-powered-by-openais-whisper (Oct 27, 2024).
53. Marcus G, Davis E. *Rebooting AI: Building Artificial Intelligence We Can Trust.* Vintage, 2019.
54. https://www.nytimes.com/2023/01/06/podcasts/transcript-ezra-klein-interviews-gary-marcus.html (Jan 6, 2023).
55. Kahneman D. *Thinking, Fast and Slow*. Macmillan, 2011.
56. Tung JYM et al. Comparison of the quality of discharge letters written by large language models and junior clinicians: single-blinded study. *J Med Internet Res* 2024; 26: e57721.
57. Chaddad A et al. Survey of explainable AI techniques in healthcare. *Sensors* 2023; 23: 634.
58. Cabral S et al. Clinical reasoning of a generative artificial intelligence model compared with physicians. *JAMA Intern Med* 2024 May 1; 184(5): 581–583.
59. Mullainathan S, Obermeyer Z. Diagnosing physician error: a machine learning approach to low-value health care. *Q J Econ* 2022; 137: 679–727.
60. Obermeyer Z. @oziadias. *X/Twitter*, https://twitter.com/oziadias/status/1513363673938468865 (Apr 11, 2022).

9 The Shotgun Marriage of Man and Machine

1. Hofling CK et al. An experimental study in nurse–physician relationships. *J Nerv Ment Dis* 1966; 143: 171–180.
2. Peadon R et al. Hierarchy and medical error: speaking up when witnessing an error. *Saf Sci* 2020; 125: 104648.
3. Dendle C et al. Why is it so hard for doctors to speak up when they see an error occurring? *Healthc Infect* 2013; 18: 72–75.
4. Marsh H. *Do No Harm: Stories of Life, Death and Brain Surgery*. Hachette UK, 2014.

5. Shem S. *The House of God*. Bantam Dell, 1978.
6. Marsh H. *Admissions: A Life in Brain Surgery*. Thomas Dunne Books, 2017.
7. Gianakos AL et al. Bullying, discrimination, harassment, sexual harassment, and the fear of retaliation during surgical residency training: a systematic review. *World J Surg* 2022; 46: 1587–1599.
8. https://www.ismp.org/resources/unresolved-disrespectful-behavior-health-care-practitioners-speak-again-part-i (2013).
9. Begeny CT et al. Sexual harassment, sexual assault and rape by colleagues in the surgical workforce, and how women and men are living different realities: observational study using NHS population-derived weights. *Br J Surg* 2023; 110: 1518–1526.
10. Schlick CJR et al. Experiences of gender discrimination and sexual harassment among residents in general surgery programs across the US. *JAMA Surg* 2021; 156: 942–952.
11. MacDonald O. Disruptive physician behavior. *QuantiaMD* (May 15, 2011).
12. https://www.england.nhs.uk/2020/02/nhs-staff-morale-improves-but-too-many-facing-abuse/ (2020).
13. Álvarez Villalobos NA et al. Prevalence and associated factors of bullying in medical residents: a systematic review and meta-analysis. *J Occup Health* 2023; 65: e12418.
14. Chadaga AR et al. Bullying in the American graduate medical education system: a national cross-sectional survey. *PLOS One* 2016; 11: e0150246.
15. Mahood SC. Medical education: beware the hidden curriculum. *Can Fam Physician* 2011; 57: 983–985.
16. Rotenstein LS et al. Prevalence of depression, depressive symptoms, and suicidal ideation among medical students: a systematic review and meta-analysis. *JAMA* 2016; 316: 2214–2236.
17. Olfson M et al. Suicide risks of health care workers in the US. *JAMA* 2023; 330: 1161–1166.
18. Kline R, Lewis D. The price of fear: estimating the financial cost of bullying and harassment to the NHS in England. *Public Money Manag* 2019; 39: 166–174.
19. https://www.nybooks.com/articles/2010/02/11/the-chess-master-and-the-computer/ (2010).
20. Cerrato P, Halamka J. *Reinventing Clinical Decision Support: Data Analytics, Artificial Intelligence, and Diagnostic Reasoning*. Taylor & Francis, 2020.
21. https://www.usertesting.com/resources/library/industry-reports/us-consumer-perceptions-ai-healthcare (2024).
22. https://www.bain.com/insights/call-the-doctor-are-patients-ready-for-generative-ai-in-healthcare-snap-chart/ (Jul 30, 2024).
23. https://www.pewresearch.org/science/2023/02/22/60-of-americans-would-be-uncomfortable-with-provider-relying-on-ai-in-their-own-health-care/ (2023).
24. Nurek M, Kostopoulou O. How the UK public views the use of diagnostic decision aids by physicians: a vignette-based experiment. *J Am Med Inform Assoc* 2023; 30: 888–898.
25. https://www.health.org.uk/publications/long-reads/ai-in-health-care-what-do-the-public-and-nhs-staff-think (Jul 31, 2024).

26. Heekin AM et al. Choosing Wisely clinical decision support adherence and associated inpatient outcomes. *Am J Manag Care* 2018; 24: 361.
27. Dave N et al. Interventions targeted at reducing diagnostic error: systematic review. *BMJ Qual Saf* 2022; 31: 297–307.
28. Staal J et al. Effect on diagnostic accuracy of cognitive reasoning tools for the workplace setting: systematic review and meta-analysis. *BMJ Qual Saf* 2022; 31: 899–910.
29. Samal L et al. Impact of electronic health records on racial and ethnic disparities in blood pressure control at US primary care visits. *Arch Intern Med* 2012; 172: 75–76.
30. Lau BD et al. Eliminating healthcare disparities via mandatory clinical decision support: the venous thromboembolism (VTE) example. *Med Care* 2015; 53: 18.
31. Lambe KA et al. Dual-process cognitive interventions to enhance diagnostic reasoning: a systematic review. *BMJ Qual Saf* 2016; 25: 808–820.
32. Kwan JL et al. Computerised clinical decision support systems and absolute improvements in care: meta-analysis of controlled clinical trials. *BMJ* 2020; 370: m3216.
33. Nanji KC et al. Overrides of medication-related clinical decision support alerts in outpatients. *J Am Med Inform Assoc* 2014; 21: 487–491.
34. Westbrook JI et al. Task errors by emergency physicians are associated with interruptions, multitasking, fatigue and working memory capacity: a prospective, direct observation study. *BMJ Qual Saf* 2018; 27: 655–663.
35. Koppel R et al. Role of computerized physician order entry systems in facilitating medication errors. *JAMA* 2005; 293: 1197–1203.
36. https://www.theverge.com/2018/9/13/17855006/apple-watch-series-4-ekg-fda-approved-vs-cleared-meaning-safe (2018).
37. Nazarian S et al. Diagnostic accuracy of smartwatches for the detection of cardiac arrhythmia: systematic review and meta-analysis. *J Med Internet Res* 2021; 23: e28974.
38. Pepplinkhuizen S et al. Accuracy and clinical relevance of the single-lead Apple Watch electrocardiogram to identify atrial fibrillation. *Cardiovasc Digit Health J* 2022; 3: S17–S22.
39. Logg JM et al. Algorithm appreciation: people prefer algorithmic to human judgment. *Organ Behav Hum Decis Process* 2019; 151: 90–103.
40. Meehl PE. *Clinical versus Statistical Prediction: A Theoretical Analysis and a Review of the Evidence*. University of Minnesota Press, 1954.
41. Burton JW et al. A systematic review of algorithm aversion in augmented decision making. *J Behav Decis Mak* 2020; 33: 220–239.
42. Dietvorst BJ et al. Algorithm aversion: people erroneously avoid algorithms after seeing them err. *J Exp Psychol Gen* 2015; 144: 114.
43. Kostopoulou O et al. Using cancer risk algorithms to improve risk estimates and referral decisions. *Commun Med* 2022; 2: 2.
44. Pálfi B et al. Algorithm-based advice taking and clinical judgement: impact of advice distance and algorithm information. *Cogn Res Princ Implic* 2022; 7: 70.
45. Agarwal N et al. *Combining Human Expertise with Artificial Intelligence: Experimental Evidence from Radiology*. Working Paper, National Bureau of Economic Research, 2023 Jul 10.

46. Jang S et al. Deep learning-based automatic detection algorithm for reducing overlooked lung cancers on chest radiographs. *Radiology* 2020; 296: 652–661.
47. Homayounieh F et al. An artificial intelligence-based chest X-ray model on human nodule detection accuracy from a multicenter study. *JAMA Netw Open* 2021; 4: e2141096.
48. Yu F et al. Heterogeneity and predictors of the effects of AI assistance on radiologists. *Nat Med* 2024; 1–13.
49. Chiang PP et al. Implementing a QCancer risk tool into general practice consultations: an exploratory study using simulated consultations with Australian general practitioners. *Br J Cancer* 2015; 112: S77–S83.
50. Meyer AN et al. Physicians' diagnostic accuracy, confidence, and resource requests: a vignette study. *JAMA Intern Med* 2013; 173: 1952–1958.
51. Khairat S et al. Reasons for physicians not adopting clinical decision support systems: critical analysis. *JMIR Med Inform* 2018; 6: e24.
52. Hautz WE et al. Diagnoses supported by a computerised diagnostic decision support system versus conventional diagnoses in emergency patients (DDX-BRO): a multicentre, multiple-period, double-blind, cluster-randomised, crossover superiority trial. *Lancet Digital Health* 2025 Feb 1; 7(2): e136-44.
53. Kostopoulou O et al. Can decision support combat incompleteness and bias in routine primary care data? *J Am Med Inform Assoc* 2021; 28: 1461–1467.
54. Kostopoulou O et al. Information search and information distortion in the diagnosis of an ambiguous presentation. *Judgm Decis Mak* 2009; 4: 408–419.
55. Tooby J, Cosmides L. The psychological foundations of culture. In: Barkow H et al. (eds) *The Adapted Mind: Evolutionary Psychology and the Generation of Culture*. Oxford University Press, 1992.
56. Sperber D, Hirschfeld LA. The cognitive foundations of cultural stability and diversity. *Trends Cogn Sci* 2004; 8: 40–46.
57. Van Strien JW et al. Testing the snake-detection hypothesis: larger early posterior negativity in humans to pictures of snakes than to pictures of other reptiles, spiders and slugs. *Front Hum Neurosci* 2014; 8: 691.
58. Liberati EG et al. What hinders the uptake of computerized decision support systems in hospitals? A qualitative study and framework for implementation. *Implement Sci* 2017; 12: 1–13.
59. Mahmud H et al. What influences algorithmic decision-making? A systematic literature review on algorithm aversion. *Technol Forecast Soc Change* 2022; 175: 121390.
60. Dennett DC. Intentional systems. *J Philos* 1971; 68: 87–106.
61. Longoni C et al. Resistance to medical artificial intelligence. *J Consum Res* 2019; 46: 629–650.
62. Berger B et al. Watch me improve—algorithm aversion and demonstrating the ability to learn. *Bus Inf Syst Eng* 2021; 63: 55–68.
63. Blease C et al. Generative artificial intelligence in primary care: an online survey of UK general practitioners. *BMJ Health Care Inform* 2024 Aug 29; 31(1): e101102.
64. Pelau C et al. Scenario-based approach to AI's agency to perform human-specific tasks. *Proc Int Conf Bus Excell* 2024; 18: 2311–2318.
65. McDuff D et al. Towards accurate differential diagnosis with large language models, http://arxiv.org/abs/2312.00164 (2023).

66. Goh E et al. Large language model influence on diagnostic reasoning: a randomized clinical trial. *JAMA Netw Open* 2024; 7: e2440969.
67. Tu T et al. Towards conversational diagnostic AI, http://arxiv.org/abs/2401.05654 (2024).
68. Singhal K et al. Toward expert-level medical question answering with large language models. *Nature Medicine* 2025 Jan 8: 1–8.
69. Bedi S et al. Testing and evaluation of health care applications of large language models: a systematic review. *JAMA* 2024 Oct 15.
70. Brodeur PG et al. Superhuman performance of a large language model on the reasoning tasks of a physician. Epub ahead of print Dec 14, 2024. doi: 10.48550/arXiv.2412.10849.
71. https://venturebeat.com/ai/openais-o3-shows-remarkable-progress-on-arc-agi-sparking-debate-on-ai-reasoning/ (2024).

10 Doctors Getting Deep

1. Osler W. *Aequanimitas*. Blakiston, 1904.
2. https://iris.who.int/handle/10665/60263 (1993).
3. Topol E. *Deep Medicine: How Artificial Intelligence Can Make Healthcare Human Again*. Hachette UK, 2019.
4. https://www.medscape.com/slideshow/2018-medical-student-report-6010086 (2018).
5. Marsh H. *Do No Harm: Stories of Life, Death and Brain Surgery*. Hachette UK, 2014.
6. Ford S et al. Can oncologists detect distress in their out-patients and how satisfied are they with their performance during bad news consultations? *Br J Cancer* 1994; 70: 767–770.
7. Pollak KI et al. Oncologist communication about emotion during visits with patients with advanced cancer. *J Clin Oncol* 2007; 25: 5748–5752.
8. Awdish R. *In Shock: How Nearly Dying Made Me a Better Intensive Care Doctor*. Random House, 2018.
9. https://www.kevinmd.com/blog/2013/09/patient-experience-outlier.html (2013).
10. Kalanithi P. *When Breath Becomes Air*. Random House, 2016.
11. Howick J et al. How empathic is your healthcare practitioner? A systematic review and meta-analysis of patient surveys. *BMC Med Educ* 2017; 17: 136.
12. https://www.gov.uk/government/publications/report-of-the-mid-staffordshire-nhs-foundation-trust-public-inquiry (2013).
13. Beach WA et al. Disclosing and responding to cancer "fears" during oncology interviews. *Soc Sci Med* 2005; 60: 893–910.
14. Ford S et al. Doctor–patient interactions in oncology. *Soc Sci Med* 1996; 42: 1511–1519.
15. Morse DS et al. Missed opportunities for interval empathy in lung cancer communication. *Arch Intern Med* 2008; 168: 1853–1858.
16. Butow PN et al. Oncologists' reactions to cancer patients' verbal cues. *Psycho-Oncol J Psychol Soc Behav Dimens Cancer* 2002; 11: 47–58.

17. Nazione S et al. Verbal social support for newly diagnosed breast cancer patients during surgical decision-making visits. *J Commun Healthc* 2016; 9: 267–278.
18. October TW et al. Characteristics of physician empathetic statements during pediatric intensive care conferences with family members: a qualitative study. *JAMA Netw Open* 2018; 1: e180351.
19. Levinson W et al. A study of patient clues and physician responses in primary care and surgical settings. *JAMA* 2000; 284: 1021–1027.
20. Vannoy SD et al. Now what should I do? Primary care physicians' responses to older adults expressing thoughts of suicide. *J Gen Intern Med* 2011; 26: 1005–1011.
21. HRH the Prince of Wales. Integrated health and post modern medicine. *J R Soc Med* 2012; 105: 496.
22. Clarke R. *Dear Life: A Doctor's Story of Love and Loss*. Little Brown UK, 2020.
23. Charon R. Narrative medicine: a model for empathy, reflection, profession, and trust. *JAMA* 2001; 286: 1897–1902.
24. Halpern J. *From Detached Concern to Empathy: Humanizing Medical Practice*. Oxford University Press, 2001.
25. Groopman JE. *The Soul of a Doctor: Harvard Medical Students Face Life and Death*. Algonquin Books, 2012.
26. Madore KP, Wagner AD. Multicosts of multitasking. *Cerebrum: The Dana Forum on Brain Science* 2019 Apr 1.
27. Haque OS, Waytz A. Why doctors should be more empathetic—but not too much more. *Sci Am* 2011 Apr; 26.
28. Jack AI et al. fMRI reveals reciprocal inhibition between social and physical cognitive domains. *NeuroImage* 2013; 66: 385–401.
29. Clarke R. *Your Life in My Hands: A Junior Doctor's Story*. Metro Publishing, 2017.
30. Haque OS, Waytz A. Dehumanization in medicine: causes, solutions, and functions. *Perspect Psychol Sci* 2012; 7: 176–186.
31. Prinz J. Against empathy. *South J Philos* 2011; 49: 214–233.
32. Bloom P. *Against Empathy: The Case for Rational Compassion*. Random House, 2017.
33. Hall JA et al. How do laypeople define empathy? *J Soc Psychol* 2020; 161: 5–24.
34. Wondra JD, Ellsworth PC. An appraisal theory of empathy and other vicarious emotional experiences. *Psychol Rev* 2015; 122: 411.
35. Halpern J. What is clinical empathy? *J Gen Intern Med* 2003; 18: 670–674.
36. Thirioux B et al. Empathy is a protective factor of burnout in physicians: new neuro-phenomenological hypotheses regarding empathy and sympathy in care relationship. *Front Psychol* 2016; 7: 763.
37. Figley CR. Compassion fatigue: toward a new understanding of the costs of caring. In: Stamm BH (ed.) *Secondary Traumatic Stress: Self-Care Issues for Clinicians, Researchers, and Educators*. Sidran Press, 1995, pp. 3–28.
38. Nielsen HG, Tulinius C. Preventing burnout among general practitioners: is there a possible route? *Educ Prim Care* 2009; 20: 353–359.
39. Greenberg DM et al. Sex and age differences in "theory of mind" across 57 countries using the English version of the "Reading the Mind in the Eyes" test. *Proc Natl Acad Sci* 2023; 120: e2022385119.

40. Gerdes KE. Empathy, sympathy, and pity: 21st-century definitions and implications for practice and research. *J Soc Serv Res* 2011; 37: 230–241.
41. Decety J, Jackson PL. A social-neuroscience perspective on empathy. *Curr Dir Psychol Sci* 2006; 15: 54–58.
42. Gerdes KE et al. Conceptualising and measuring empathy. *Br J Soc Work* 2010; 40: 2326–2343.
43. Batson CD et al. Empathy and attitudes: can feeling for a member of a stigmatized group improve feelings toward the group? *J Pers Soc Psychol* 1997; 72: 105.
44. Hein G et al. Neural responses to ingroup and outgroup members' suffering predict individual differences in costly helping. *Neuron* 2010; 68: 149–160.
45. Xu X et al. Do you feel my pain? Racial group membership modulates empathic neural responses. *J Neurosci* 2009; 29: 8525–8529.
46. Avenanti A et al. Racial bias reduces empathic sensorimotor resonance with other-race pain. *Curr Biol* 2010; 20: 1018–1022.
47. Decety J et al. Empathy, justice, and moral behavior. *AJOB Neurosci* 2015; 6: 3–14.
48. Batson CD et al. Immorality from empathy-induced altruism: when compassion and justice conflict. *J Pers Soc Psychol* 1995; 68: 1042.
49. Kay A. *This Is Going to Hurt: Secret Diaries of a Junior Doctor*. Pan Macmillan, 2017.
50. Dunbar RI. Neocortex size as a constraint on group size in primates. *J Hum Evol* 1992; 22: 469–493.
51. Dunbar RI. Do online social media cut through the constraints that limit the size of offline social networks? *R Soc Open Sci* 2016; 3: 150292.
52. Arnaboldi V et al. Egocentric online social networks: analysis of key features and prediction of tie strength in Facebook. *Comput Commun* 2013; 36: 1130–1144.
53. Sutcliffe A et al. Relationships and the social brain: integrating psychological and evolutionary perspectives. *Br J Psychol* 2012; 103: 149–168.
54. Beck CT. Secondary traumatic stress in nurses: a systematic review. *Arch Psychiatr Nurs* 2011; 25: 1–10.
55. Potter P et al. Compassion fatigue and burnout: prevalence among oncology nurses. *Clin J Oncol Nurs* 2010; 14: E56–62.
56. Rossi A et al. Burnout, compassion fatigue, and compassion satisfaction among staff in community-based mental health services. *Psychiatry Res* 2012; 200: 933–938.
57. Buunk BP et al. Perceived reciprocity, social support, and stress at work: the role of exchange and communal orientation. *J Pers Soc Psychol* 1993; 65: 801.
58. Cosmides L, Tooby J. Adaptations for reasoning about social exchange. In: Buss DM (ed.) *The Handbook of Evolutionary Psychology*. Wiley, 2015: 1–44.

11 Humanizing Healthcare without Doctors

1. Blease C et al. Computerization and the future of primary care: a survey of general practitioners in the UK. *PLOS One* 2018; 13: e0207418.
2. Blease C et al. Artificial intelligence and the future of primary care: exploratory qualitative study of UK general practitioners' views. *J Med Internet Res* 2019; 21: e12802.

3. Doraiswamy PM et al. Artificial intelligence and the future of psychiatry: insights from a global physician survey. *Artif Intell Med* 2020; 102: 101753.
4. Blease C et al. Artificial intelligence and the future of psychiatry: qualitative findings from a global physician survey. *Digit Health* 2020; 6: doi: 10.1177/2055207620968355.
5. https://www.siliconrepublic.com/start-ups/dublin-cork-belfast-europe-tech-cities (2021).
6. Blease C et al. Computerization of the work of general practitioners: mixed methods survey of final-year medical students in Ireland. *JMIR Med Educ* 2023; 9: e42639.
7. Blease C et al. Generative artificial intelligence in medicine: a mixed methods survey of UK general practitioners. Unpubl Manuscr, under review.
8. Topol EJ. Machines and empathy in medicine. *The Lancet* 2023; 402: 1411.
9. Topol E. *Deep Medicine: How Artificial Intelligence Can Make Healthcare Human Again.* Hachette UK, 2019.
10. https://www.nytimes.com/2002/12/08/magazine/the-way-we-live-now-12-8-02-the-healing-paradox.html (Dec 8, 2002, accessed Jun 9, 2023).
11. Charon R. Narrative medicine: a model for empathy, reflection, profession, and trust. *JAMA* 2001; 286: 1897–1902.
12. Milota MM et al. Narrative medicine as a medical education tool: a systematic review. *Med Teach* 2019; 41: 802–810.
13. Gerger H et al. Lay perspectives on empathy in patient–physician communication: an online experimental study. *Health Commun* 2023; 1–10.
14. Hall JA et al. What is clinical empathy? Perspectives of community members, university students, cancer patients, and physicians. *Patient Educ Couns* 2021; 104: 1237–1245.
15. Beach MC et al. Is physician self-disclosure related to patient evaluation of office visits? *J Gen Intern Med* 2004; 19: 905–910.
16. Steitz BD et al. Perspectives of patients about immediate access to test results through an online patient portal. *JAMA Netw Open* 2023; 6: e233572.
17. Kelley JM et al. The influence of the patient–clinician relationship on healthcare outcomes: a systematic review and meta-analysis of randomized controlled trials. *PLOS One* 2014; 9: e94207.
18. Scheder-Bieschin J et al. Improving emergency department patient–physician conversation through an artificial intelligence symptom-taking tool: mixed methods pilot observational study. *JMIR Form Res* 2022; 6: e28199.
19. Webb JJ. Proof of concept: using ChatGPT to teach emergency physicians how to break bad news. *Cureus* 2023 May; 15(5).
20. Tanana MJ et al. Development and evaluation of ClientBot: patient-like conversational agent to train basic counseling skills. *J Med Internet Res* 2019; 21: e12529.
21. Sharma A et al. Human–AI collaboration enables more empathic conversations in text-based peer-to-peer mental health support. *Nat Mach Intell* 2023; 5: 46–57.
22. Blease C et al. Generative language models and open notes: exploring the promise and limitations. *JMIR Med Educ* 2024; 10: e51183.
23. Kharko A et al. Generative artificial intelligence writing open notes: a mixed methods assessment of the functionality of GPT 3.5 and GPT 4.0. *Digit Health* 2024; doi: 10:1177/20552076241291384.

24. Ayers JW et al. Comparing physician and artificial intelligence chatbot responses to patient questions posted to a public social media forum. *JAMA Intern Med* 2023; 183: 589–596.
25. Elyoseph Z et al. ChatGPT outperforms humans in emotional awareness evaluations. *Front Psychol* 2023; 14: 1199058.
26. Sufyan NS et al. Artificial intelligence and social intelligence: preliminary comparison study between AI models and psychologists. *Front Psychol* 2024; 15: 1353022.
27. Lovens P-F. Sans ces conversations avec le chatbot Eliza, mon mari serait toujours là. *La Libre*, Mar 28, 2023.
28. Sharp G et al. Ethical challenges in AI approaches to eating disorders. *Journal of Medical Internet Research* 2023; 25: e50696.
29. https://www.nbcnews.com/tech/internet/chatgpt-ai-experiment-mental-health-tech-app-koko-rcna65110 (Jan 14, 2023).
30. Yin Y et al. AI can help people feel heard, but an AI label diminishes this impact. *Proc Natl Acad Sci* 2024; 121: e2319112121.
31. Siddals S et al. "It happened to be the perfect thing": experiences of generative AI chatbots for mental health. *npj Ment Health Res* 2024; 3: 48.
32. https://www.bbc.co.uk/news/articles/cy7g45g2nxno (Feb 12, 2025).
33. Barkow J. *Darwin, Sex, and Status: Biological Approaches to Mind and Culture*. University of Toronto Press, 1989, p. 118.
34. Kanazawa S. Bowling with our imaginary friends. *Evol Hum Behav* 2002; 23: 167–171.
35. Gardner WL, Knowles ML. Love makes you real: favorite television characters are perceived as "real" in a social facilitation paradigm. *Soc Cogn* 2008; 26: 156–168.
36. Mar RA et al. Exploring the link between reading fiction and empathy: ruling out individual differences and examining outcomes. *Communications* 2009; 34: 407–428.
37. Epley N et al. Motivated mind perception: treating pets as people and people as animals. In: Gervais SJ (ed.) *Objectification and (De)Humanization*. Springer New York, 2013, pp. 127–152.
38. Govindarajan V, Ramamurti R. Transforming health care from the ground up. *Harv Bus Rev* 2018; 96–104.
39. Bickerdike L et al. Social prescribing: less rhetoric and more reality. A systematic review of the evidence. *BMJ Open* 2017; 7: e013384.
40. Kiely B et al. Effect of social prescribing link workers on health outcomes and costs for adults in primary care and community settings: a systematic review. *BMJ Open* 2022; 12: e062951.
41. Mercer SW et al. Effectiveness of community-links practitioners in areas of high socioeconomic deprivation. *Ann Fam Med* 2019; 17: 518–525.
42. https://educationdata.org/average-cost-of-medical-school (2024).
43. Youngclaus J, Fresne J. *Physician Education Debt and the Cost to Attend Medical School: 2020 Update*. Association of American Medical Colleges, 2020.
44. Steven K et al. Fair access to medicine? Retrospective analysis of UK medical schools application data 2009–2012 using three measures of socioeconomic status. *BMC Med Educ* 2016; 16: 11.

45. Arulampalam W et al. Factors affecting the probability of first-year medical student dropout in the UK: a logistic analysis for the intake cohorts of 1980–92. *Med Educ* 2004; 38: 492–503.
46. Choi KJ et al. Characteristics of medical students with physician relatives: a national study. *MedEdPublish* 2018; 7.
47. Polyakova M et al. Does medicine run in the family—evidence from three generations of physicians in Sweden: retrospective observational study. *BMJ* 2020; 371: m4453.
48. Dunbar RI. The social brain hypothesis and its implications for social evolution. *Ann Hum Biol* 2009; 36: 562–572.
49. Dunbar RI. Coevolution of neocortical size, group size and language in humans. *Behav Brain Sci* 1993; 16: 681–694.
50. Weiner DB. The apprenticeship of Philippe Pinel: a new document, "observations of Citizen Pussin on the insane." *Am J Psychiatry* 1979 Sep 1; 136(9): 1128–1134.
51. https://www.pewresearch.org/internet/2020/02/06/the-virtues-and-downsides-of-online-dating/ (2020).
52. Mulvale G et al. Integrating mental health peer support in clinical settings: lessons from Canada and Norway. *Healthc Manage Forum* 2019; 32: 68–72.

Conclusion: Leaving the Appointment

1. Henderson R. *Troubled: A Memoir of Foster Care, Family, and Social Class.* Simon & Schuster, 2025.
2. https://www.itgovernance.eu/blog/en/data-breaches-and-cyber-attacks-in-europe-in-february-2024-50884709-records-breached (2024).
3. https://eu.usatoday.com/story/news/health/2024/02/18/health-data-breaches-hit-new-record-2023/72507651007/ (2024).
4. https://www.bbc.co.uk/news/articles/c9ww90j9dj8o (Jun 21, 2024).
5. https://www.wsj.com/articles/google-s-secret-project-nightingale-gathers-personal-health-data-on-millions-of-americans-11573496790 (Nov 11, 2019).
6. https://www.theguardian.com/technology/2019/nov/12/google-medical-data-project-nightingale-secret-transfer-us-health-information (Nov 12, 2019).
7. https://themarkup.org/privacy/2022/12/13/out-of-control-dozens-of-telehealth-startups-sent-sensitive-health-information-to-big-tech-companies (Dec 13, 2022).
8. Blease C. OpenAI meets open notes: surveillance capitalism, patient privacy and online record access. *J Med Ethics* 2024; 50: 84–89.
9. Zuboff S. *The Age of Surveillance Capitalism: The Fight for a Human Future at the New Frontier of Power.* Profile Books, 2019.
10. https://www.pewresearch.org/internet/2019/11/15/americans-and-privacy-concerned-confused-and-feeling-lack-of-control-over-their-personal-information/ (2019).
11. Coiera E. The last mile: where artificial intelligence meets reality. *J Med Internet Res* 2019; 21: e16323.
12. https://www.techdirt.com/2010/04/09/institutions-will-seek-to-preserve-the-problem-for-which-they-are-the-solution/ (2010).

13. Blease C et al. Computerization and the future of primary care: a survey of general practitioners in the UK. *PLOS One* 2018; 13: e0207418.
14. Blease C et al. Artificial intelligence and the future of primary care: exploratory qualitative study of UK general practitioners' views. *J Med Internet Res* 2019; 21: e12802.
15. Blease C et al. Artificial intelligence and the future of psychiatry: qualitative findings from a global physician survey. *Digit Health* 2020; 6: 10.1177/2055207 620968355.
16. Doraiswamy PM et al. Artificial intelligence and the future of psychiatry: insights from a global physician survey. *Artif Intell Med* 2020; 102: 101753.
17. Blease C et al. Machine learning in medical education: a survey of the experiences and opinions of medical students in Ireland. *BMJ Health Care Inform* 2022; 29(1): e100480.
18. Blease C et al. Computerization of the work of general practitioners: mixed methods survey of final-year medical students in Ireland. *JMIR Med Educ* 2023; 9: e42639.
19. Blease C et al. Psychiatrists' experiences and opinions of generative artificial intelligence in mental healthcare: an online mixed methods survey. *Psychiatry Res* 2024; 333: 115724.
20. Blease C et al. Generative artificial intelligence in primary care: an online survey of UK general practitioners. *BMJ Health Care Inform* 2024 Aug 29; 31(1): e101102.
21. https://azure.microsoft.com/en-us/blog/introducing-gpt4-in-azure-openai-service/ (Mar 21, 2023).
22. Wootton D. *Bad Medicine: Doctors Doing Harm since Hippocrates*. Oxford University Press, 2007.
23. Blease C et al. US primary care in 2029: a Delphi survey on the impact of machine learning. *PLOS One* 2020; 15: e0239947.
24. Blease C, Torous J. ChatGPT and mental healthcare: balancing benefits with risks of harms. *BMJ Ment Health* 2023; 26: e300884.
25. Dos Santos DP et al. Medical students' attitude towards artificial intelligence: a multicentre survey. *Eur Radiol* 2019; 29: 1640–1646.
26. https://lcme.org/publications/#DCI (Mar 2024).
27. https://topol.hee.nhs.uk/ (2019).
28. https://www.gmc-uk.org/education/standards-guidance-and-curricula/standards-and-outcomes/outcomes-for-graduates (2020).
29. Susskind RE, Susskind D. *The Future of the Professions: How Technology Will Transform the Work of Human Experts*. Oxford University Press, 2015.
30. https://www.kff.org/health-misinformation-and-trust/poll-finding/kff-health-misinformation-tracking-poll-artificial-intelligence-and-health-information/ (Aug 15, 2024).

Acknowledgments

This book would not have been possible without support, advice, and discussions with many people. To each one of you, a warm thank you: Helen Atherton, Grayson Baird, Michael H. Bernstein, Catherine Blease, Nicholas Blease, Rosemary Blease, Stephen Blease, Victor Blease, Dave deBronkart ('ePatient Dave'), Gail Davidge, Berkeley Dietvorst, Emma Doble, Brian Donnelly, Leo Druart, Robin Dunbar, Rushika Fernandopulle, Jens Gaab, Keith Geraghty, James Guszcza, Maria Hägglund, Josefin Hagström, Kathryn Hall, Amir Hannan, Wolf Hautz, Jenny Holland, Anne Holloway, William Howe, Joanne (Jo) Hunt, Lisa Iezzoni, Ray Jones, Juliane Kämmer, Karla Kendrick, Anna Kharko, Olga Kostopoulou, Kathryn Lambe, Liliana Laranjo, E. Thomas Lawson, Jennifer (Jen) Lawson, Ruth Lawson, Sonya Lawson, Richard Lehman, Cosima Locher, Jennifer Logg, Courtney Lyles, Maxine Mackintosh, Brian McMillan, Kenneth D. Mandl, Arjun (Raj) Manrai, Jason Maude, Raj Mehta, Matthew (Matt) Might, Leanne Morgan, Joel Mort, Cathy O'Neil, Ed Orgill, Hanife Rexhepi, Marelle Rice, Sara Riggare, Adam Rodman, Jorge Rodriguez, Rachel Schlund, Ryan Schuetzler, Henry Scowcroft, Richard Sills, Hardeep Singh, John Torous, Andrew Turner, Allen Wenner, Carl Whyte, Laurie Wilson-Moore, and Laura Zwaan. Two supporters and interviewees for this book, who have since passed away, deserve a special mention: Jerome Barkow and Ingrid Brindle.

A big thank you is owed to my agent Caroline Hardman who was an early believer and champion of the project, as was Sarah Levitt, who steered its initial direction. My colleagues and friends at Yale University Press have been patient throughout the process. Heather McCallum championed the central idea and encouraged me from the get-go, and Katie Urquhart, Heather Nathan, and Chloe Foster have shown inspiring enthusiasm and commitment to this project. Rachael Lonsdale and Sophie Richmond were eagle-eyed editors.

Gratitude is due to the following people who read chapters and offered feedback: Robin Dunbar, the Lawson Family (Jen, Tom, Ruth, and Sonya), and John Torous. The following read and offered constructive advice on the entire manuscript: Catherine Blease, Richard Lehman, Kenneth D. Mandl, Joel Mort, and Carl Whyte.

ACKNOWLEDGMENTS

The book was written and revised in different locations over many years, from Belfast, Northern Ireland to Boston in the US and latterly, Uppsala in Sweden. Before the book was a formal proposal, Tom Lawson, whom I met in Belfast, was my first supporter. On multiple occasions I was a guest of the Lawsons in their beautiful home in Kalamazoo, Michigan, even accompanying them on a family holiday, where we explored science, medicine, and patient experiences. I am grateful for their kindness, candor, and friendship.

At Harvard Medical School, physicians and informaticians Kenneth D. Mandl, Adam Rodman, and John Torous deserve mention for their friendship, and support. I also benefited from countless stimulating conversations with medical doctors at Beth Israel Deaconess Medical Center, Boston. There I was privileged to spend several years with OpenNotes—including Tom Delbanco, Catherine DesRoches, Jan Walker, Sigall Bell, Liz Salmi, Kendall Harcourt, and Paola Miralles—a team dedicated to patient-centric healthcare. Huge thanks are also owed to my colleagues at the Participatory eHealth and Health Data Research Group at Uppsala University in Sweden. A special note goes to Maria Hägglund, Anna Kharko, Sara Riggare, and Josefin Hagström for their kindness over the past few years.

Tony and June Matthews extended their generous hospitality to Henry McDonald (my late partner) and me, allowing us to spend truly blissful days writing together in their holiday home in Murcia, Spain. The book was also revised and rewritten in the home of my twin sister Catherine Blease and her partner Martin (Marty) Hay. Their generosity, unassuming support, and home cooking meant everything. This book would not exist without them.

Patients have been at the forefront of my mind and their plight kept me motivated. I am deeply grateful to Jen and Tom Lawson, Keris Myrick, Keith Geraghty, Jo Hunt, and the unnamed patients for sharing their stories. Throughout the writing my siblings were, and remain, a beacon of motivation: I am acutely aware of the health, medical, and social experiences of my eldest brother Stephen and sister Catherine, who live with myotonic muscular dystrophy and embody personal fortitude and resilience.

In February 2023 I lost my beloved partner Henry to cancer. Henry was a well-renowned, highly talented journalist and author of a dozen books. We discussed this project countless times and, with the deepest support, he read drafts of early chapters. It is one of life's cruel ironies that the project would take on a personal aspect. I witnessed Henry confront his illness and fate in ways that were stoic and heroic. Six months after Henry died, I lost my father Victor to dementia. He was a source of unquenchable enthusiasm and encouragement—even in the final months of his life.

This book is dedicated to their memory, to my mother, Rosemary, whose unwavering love and steadfast support have been my anchor, and to my dearest, lifelong companion—my twin, Catherine. I hope it makes them proud. Special thanks are also due to Carl Whyte, Stephanie, and Adeline, who came into my life during the final stages of this book, filling it with promise.

Finally, this book is dedicated to my readers—those who have been, are, or will be patients.

Charlotte Blease
February 22, 2025

Index

Academic Medicine journal 91
access
   to AI tools 261–2
   to doctors 2, 5, 6–7, 53–6, 64–5, 260
   equality of 54, 261–2, 276
   to health facilities 47–50
   to health records 23–4, 92–3
   rural areas 48–9, 56, 67–8
   see also telemedicine
Access to Health Records Act (1990 UK) 23
Affordable Care Act (2010 US) 45, 46
ageism 153
aging 4, 131
   and doctors 130–1
   medical training in 134–5
AI (artificial intelligence) 8–9, 276
   "agent AI" 32
   algorithmic decision-making 259
   ambient 184
   and choice of human collaborator 209–11
   and clinical decision support systems (CDSS) 197–9
   and clinical reasoning 185–6
   compared with doctors 185–7
   connectionist 106
   and data security 261–2
   to detect dementia 119
   doctors' relationship with 195–201
   education in 268–9
   expert systems 104–5
   explainable 185–6
   "generative AI" 29
   identification of race 180
   and imaging 125, 176, 201
   lack of transparency 184–5, 249–50
   limitations of 177–8, 184–5, 187
   and medical images 178–9, 180, 201
   need for empirical research 33–4
   and paperwork 267
   possibility of paywalls 261
   possible medical roles 270–73
   and potential to reduce bias 178–82
   and sentience 247
   and training biases 174–5, 176–7, 248
   training sets 167
   see also chatbots; OpenAI
"AI Winters" 34
AIDS 230–1, 248
Alabama, University of 173, 272, 273
Album, Dag 150–1
alcohol disclosures 112, 243–4, 263
"alert fatigue" 198
algorithmic appreciation 200, 201
algorithmic aversion 200–1
algorithmic decision-making 259

INDEX

algorithmic discrimination 174–5, 176–7, 259, 271
allergies 41, 99
Alphabet 24
AlphaZero chess AI 210
Althoff, Tim 244
Amara's Law 34
Ambient Experience app 267
American Association of Medical Colleges 217
American Heart Association 136
American Hospital Association 24
American Journal of Public Health 46–7
American Medical Association (AMA)
    Ethics Code 162–3
    exclusion of Black physicians 55
    lobbying 24
    on telemedicine 72
American Time Survey Data 50
Americans with Disabilities Act (1990) 51
AMIE, LLM chatbot 106
Anastos-Wallen, Rebecca E. 64, 73
Ann Arbor, Michigan 48
Anthropic, Claude chatbot 29, 106, 176
anthropomorphism 110–12, 204
    applied to AI 204–5
    and "design stance" 205–6
    and "intentional stance" 205, 206–7
antibiotics, prescriptions 125
Apple Watch, detection of AFib 188–90, 199, 205
appointments
    time allowed 103
    updating records 103–4
Armstrong, Izzy 39
Ascension, Catholic health organization 263
Astroten experiment 191
atrial fibrillation, Apple Watch diagnosis 188–90, 199, 205
Australia 82, 137, 221, 226
automation 31–2, 104
avatars 112
Awdish, Rana 226, 235
    In Shock 218
"Ayan, Dr. Emir" (pseudonym) 215–16, 232, 234, 235

Ayers, John 247–8
Azure platform, Epic Systems 267

Bachman, John 108
Bacon, Francis 33–4
Bangladesh 215
Barkow, Jerome 250–1
Batson, Daniel 230–1, 233, 248
Beach, Mary 241–2
Beam, Andrew 171
bedside manner 216, 221–3, 224
    doctors' view of 237–8
    see also empathy
Bender, Emily 183
benign prostatic hyperplasia (BPH) 79
Benjenk, Ivy 69
Berkshire Hathaway 83
Bernstein, Carl 153
Beth Israel Deaconess Medical Center 61–2, 92, 166, 274
    Digital Psychiatry 116–17
Beyoncé 28
bias
    in AI training 174–5, 176–7, 248
    and anti-bias training 162–4
    availability 158
    confirmation 158
    cultural (in AI) 246
    influence on empathy 230–2, 260
    male research 133–4
    Med-PaLM-2 and 209
    potential of AI to reduce 178–82, 261
    and stereotypes 158–60
    see also discrimination; prejudice
BigTech, data tracking 263–5
Bing 263
Bing AI 168
Biomedical Data Translator Project 272–3
bipolar disorder, voice features 118
Blease, Victor 11
"Blenderbot 3" chatbot 247
Blockbuster 31
blood pressure readings 114–15
Bloom, Paul, *Against Empathy* 228
Bloomberg 29
BMJ (*British Medical Journal*) 20
Boissy, Adrienne 72

INDEX

Boston Center for Independent Living 62
Boston, Mass. 47, 57–8
breast cancer 134, 180
Brigham Young University 64, 112, 182
British Medical Association 24, 86, 87
Buckley, William F. 88
Buffett, Warren 83–4, 88
bullying 193–5
Burns, George 242
Butler, Robert 153

Cambridge University, Institute of Public Health 140
cancer, discriminatory treatment 148–9
cancer diagnoses, and doctors' empathy 221
cancer screening, referrals 125
Care Quality Commission, on complaints 97
Cedars-Sinai Health System, Los Angeles 197
Centers for Medicare and Medicaid Services 26
Chaiyachati, Krisda 64, 73, 207
Charles III, King, as Prince of Wales 223, 225
Charon, Rita 224, 229, 239–40
chatbots
 and consistency 169–70
 and cultural competence 246
 designers 271
 and diagnostic accuracy 108–9, 181, 186, 203, 268
 and hallucinations 183, 184, 185, 269
 and hidden data patterns 171–3
 information parameters 168–9
 linguistic competence 175
 and Med-PaLM-2 209
 patients' use of 35, 93–5, 113, 248–51, 274–5
 and perception of comfort 250–1
 possible autonomy of 275
 prompt-engineering 169–70
 and racial bias 176–7
 reliance on 113
 risks 249
 success in medical exams 165–6, 168
 and training for empathy 243–6, 247–9
 see also ChatGPT; LLMs (large language model) chatbots; OpenAI
ChatGPT 9, 29, 106
 and American medical exams 165–6
 and AskDocs responses 247–8
 compared with OpenAI systems 210–11
 and detection of dementia 119
 and diagnosis of medical cases 166
 and difficult conversations 243–4
 doctors' use of 34–5, 94, 207, 208–9
 patients' use of 93
 and rare diseases 171, 173
 rewritten health records 245–6
 web-text training 170–1
ChatGPT-3.5 176, 184
 Emotional Awareness Scale 248
ChatGPT-4 176, 184
 clinical reasoning 186
 compared with psychologists 248
 discharge letters 185
 and Japanese medical exam 165, 168
Checker Cabs 43
Cheng, Joey 83–5, 87–8
chess
 AlphaZero and 210
 ZackS team 195–6, 198, 210, 273
Chigaco, University of 77, 148, 200
China 175, 215, 250
Christensen, Clayton, *The Innovator's Dilemma* 30–1
chronic conditions
 dismissed 151
 monitoring 114–15
chronic fatigue syndrome (ME) 144–5
Cifu, Adam 139–40
Civil Rights Act (1964 US) 55
Clarke, Rachel 227, 232–3
 Dear Life 95, 223
class
 and access to healthcare 146–7
 and discrimination 145–7, 232, 254–5
 see also race and ethnicity

INDEX

Claude chatbot 29, 106, 176
ClientBot, and empathy 243
clinical binocularity 225–6
clinical decision support systems (CDSS) 197–9
   Isabel tool 202, 203
clinical notes
   and AI 267
   prejudice in 148, 152
   see also health records
clinical research and trials
   and AI efficiency 173
   changing treatments and practices 139–40
   male bias 133–4
   and older patients 135
   and race or ethnicity 135–7
   rare diseases 137–9
   slow adoption of 140
   see also medical training
ClinicalTrials.gov 135
COBRA 45
Cochrane GPT study 115, 183, 185
communication
   and chatbots 274
   electronic 20–1
   Instant Medical History 101–3, 104
   and patient passivity 150
   patient–doctor 147, 148–50, 243–4
   and racial concordance 149
   see also chatbots; computers; empathy; telemedicine
compassion 229–30, 240
   chatbot training for 243–4
   see also empathy
compassion fatigue 235–6
complaints against doctors 26, 96–7
computers
   early use of 27–8, 99–101
   human responses to 109
   as interviewers 101–4
   patients' honesty with 107–8, 114
   see also AI; internet; telemedicine
"Connell, Emma" 259
Cook, Tim, Apple 123
Cornell University 177
Coronation Street 39

COVID-19 pandemic 3, 46
   deaths of adults with disabilities 53–4
   Disability Rights Monitor Dashboard 152
   effect on medical training 128
   misdiagnosis 158
   racial disparities (US) 54, 136
   and telemedicine 57–8, 60, 61, 66
Cramton, Steven 195
cystic fibrosis 138

The Daily Mail 29
The Daily Star 110
data
   breaches 262–5
   control of 261
   cyber-attacks 262
   misuse of 264
   unauthorized access and sale of 263–4
data entry 103, 203, 216, 267
Data Protection Act (1998 UK) 23
Decety, Jean 232
DeepSeek chatbot 29, 106, 248–9
DeepSeek-R1 32, 186
deference, to doctors 192–3, 260
   and bullying 193–5
   dominination and submission 87–8, 89
   and embarrassment about symptoms 79–83, 107
   and prestige of doctors 83–9
   problem of 77–98
Delbanco, Tom 92
dementia
   AI detection of 119
   in doctors 131
Denmark 108–9, 118
Dennett, Daniel 205–6
   design stance 205
   intentional stance 205
Devi, Gayatri 131
diabetes 152
diagnosis
   biases in 158–60
   ChatGPT tests 166
   delays for women 133

INDEX

dismissive of symptoms 143–5
  of rare disorders 137–9, 171, 173, 273
  by video telemedicine 71–2
diagnostic errors 1–2, 260
  and doctors' burnout 5
  by newly trained physicians 127–8
  research into 16–18
Diamandis, Pete, XPRIZE Foundation 28
Dietvorst, Berkeley 200–1, 206
digital capital 68, 261
digital divides 67–8, 260–61
  see also race and ethnicity
"digital phenotyping" 116–17
  see also Apple Watch; smartphones
disability
  and access to healthcare 61–3
  deaths from COVID-19 53–4
  doctors' perceptions of patients with 153
  and poor hospital design 50–3
  poverty (US) 46
  and telemedicine 62–3
  and transport barriers 49–50
  and treatment failures of other diseases 152
discrimination
  against disabled patients 152–3
  against sexual minorities 155–6
  algorithmic 174–5, 176–7
  and body weight 154–5
  and low incomes 49–50, 146–7, 149, 254
  see also bias; class; disability; race and ethnicity
diseases
  causal links to risks 180
  hierarchy of 150–3
Doctor.com, survey 59
doctors
  annual awards 123–4
  on "the art of medicine" 156–7
  attire 87
  burnout, depression and suicide 3–4, 5, 226–8, 242
  claim of superiority 18–19, 238–9
  and clinical decision support systems (CDSS) 197–9
  and cognitive decline 130–1
  compared with AI 185–7
  competence compared with chatbots 170–1
  conservatism 19–21, 32–3, 64–6, 265–6
  and defensive medicine 5–6
  disparagement by 90–2
  domination of dialogue 89–90
  and empathy 215–18, 224–5
  experience of illness 218–20
  and gift-giving 85–6
  inconsistency of judgments 125
  lack of feedback 141–2
  motivations 215, 217
  and objectivity 10, 12
  over-confidence 14–15
  and paperwork 267
  and patient access to records 24
  and patient health monitoring 116–18
  patients' fear of 95–7
  and personal emotions 241–2
  possible changing role of 270–71
  relations with nurses 190–5
  reluctance to use AI 195–201, 202–7, 266–8
  resistance to telemedicine 64–6
  shortage of 4–5
  socio-economic background of 254–5
  thinking systems and bias 160–2, 172, 179
  unchallenged 26, 191–2
  use of ChatGPT 34–5, 94, 207, 208–9
  workloads 215–16, 238–9
  see also medical profession; patients
domestic violence 108
  AI prediction of 119
drug abuse
  overdose deaths (US) 60
  and telemedicine 60–1
drugs, for older patients 135
Duke Medical School, survey 237

INDEX

Dunbar, Robin 234–5
  Dunbar's number 235, 255
DVD rentals 31

eating disorders, female bias 134
eConsult (UK) 105
Ehlers-Danlos syndrome (EDS) 39, 40–2, 137, 138
Eichstaedt, Johannes 119
ELIZA, early computer program 100
Elsevier 132
email, consultations using 108–9
embarrassment, and disclosure of symptoms 79–83, 107
Emory University 161, 181
"Empathetics" course 243
empathy 3, 215–36, 260
  adult primary care 222
  bias and 230–2, 260
  and cancer diagnoses 221
  chatbots 238, 243–6, 247–9, 261
  and clinical binocularity 225–6
  cognitive 229, 240–1, 248
  and compassion 229–30, 240–1
  and compassion fatigue 235–6
  definitions 228–9
  effect of narratives on 230–1, 233–4
  elderly patients 222–3
  emotional 229, 240–1
  human intentions and 250
  limits of 234–5
  medical training 217, 223–4, 243–4
  patients' perception of 220–1, 239–40
  and pediatric intensive care 221–2
  and peer support 256–7
  and personal experience of illness 218–20
  problems with 228
  and prosocial behavior 229
  and race and ethnicity 231
  and social justice 232–4
  see also bedside manner; communication
endometriosis 133, 171
Epic Systems, Azure platform 267

Equality Act (2010 UK) 51
Ericsson, Anders 125–6, 141
Eriksen, Alexander 166
EU Commission, and telemedicine 66
European Medicines Agency 135
European Union 2, 46, 60, 137
  access to care 48, 49
  Denied the Right to Health report (2023) 53
  "Digital Decade" program (2021) 69
  and disability 49, 51, 53
evolution 7–8, 15, 91
  and cognitive modules 204–5
evolutionary psychology 84, 154, 160, 250
expertise
  aging doctors 130–1
  failure of feedback 141–2
  female versus male doctors 129–30, 155
  and practice 123, 126–7, 141–2
  see also diagnostic errors

face detection, in objects 110–11
Facebook 119
facial recognition software, limitations 175
falls, risk of 44
Faust, Jeremy 182
fax machines 20–1
FDA (Food and Drug Administration) 29, 138, 199
Feathers, Todd 263
Fernandopulle, Rushika 251–4, 255–6, 257
Florida, University of 89
Florida State University 125
Fondrie-Teitler, Simon 263
Ford, Henry II 83, 84
Francis, Robert, QC, Inquiry 220, 223
fraud 26
Freestyle Chess Tournament 195, 196
Frosch, David 95–6

Gaga, Lady 123
Gates, Bill 165
Gebru, Timnit 183
Gemini chatbot 29, 106, 176

INDEX

gender differences
 chatbots and 176
 pain 133
 symptoms 132–3
 see also women
General Medical Council (UK) 269
"generative AI" 29, 165
 and hallucinations 182–3
 see also ChatGPT
Gerada, Clare 96
Geraghty, Keith 144–5, 151, 173
Gerger, Heike 240–1
geriatric medicine, lack of training 134–5
al-Gharbi, Musa 163
Gichoya, Judy Wawira 181
Goffman, Erving 80, 169
Goldman, Brian, *The Secret Language of Doctors* 91
Google 93–4
 and access to patients' data 263
 and AMIE 106
 Deep Mind 208
 Research 208
 see also Gemini
Gordon, Howard 149–50
GPT (Generative Pre-trained Transformer model) 165
 see also ChatGPT; OpenAI
The Guardian 11, 263
Guha, Neel 171

Hägglund, Maria 93
HAILEY chatbot 244–5
Halamka, John 196
Hall, Judith 241
Halpern, Jodi 229
 From Detached Concern to Empathy 224
Haque, Omar Sultan 226–7
Harvard, T.H. Chan School of Public Health 59
Harvard Medical School 10, 62, 119, 128, 141, 251
 and AI 172–3, 178, 184, 201
 care quality study 124
 and doctors' competence 130
 The Soul of a Doctor 225
Harvard University 83

Haze, Tatsuya 165, 169
"healing buzz" 27
Health Affairs journal 95
"health coaches" 252–4, 272
 doctors' objections to 257
 and peer support 256–7
Health and Human Services Office of Civil Rights 262
health insurance 2, 24, 46–7, 49, 149, 159
 Medicaid 45, 147
 Medicare 26, 45, 61, 129
Health Insurance Portability and Accountability Act (1996 US) 23
health monitoring 114–16
 and privacy concerns 120
 smartphones for 115–19
health records
 access to 23–4, 92–3
 ChatGPT rewritten 245–6
 updating 103–4
 see also clinical notes
healthcare engineers, AI 271
Healthcare IT News 72
healthcare systems
 and chronic illness 47
 costs of 45
 failings of 276
 gatekeepers 145–6
 inconsistency 124–5
 peer support 256–7
 prejudice in 145 50
 pressure on 5, 6–7
 travelling distance and costs 48–50
heart attacks, diagnoses of 186–7
heart failure, symptoms 132–3
Heider, Fritz 111
Henderson, Rob 259
Henrich, Joseph 83–5, 87–8
Herold, Don 141
Hinton, Geoffrey 28
Hippocrates 80
Hoffman, Kelly 137
hospitals
 parking fees 50
 poor design 50–3

INDEX

House, Dr Gregory 227
"Human + AI" responses 244–5

IBM, Watson machine 28
Iezzoni, Lisa 61–3, 65, 153
 More than Ramps 62
Illinois, University of 149
Imperial College London 201, 203
impression management 80
Inflection, Pi chatbot 248–9
Infrastructure Investment and Jobs Act (US) 68–9
Instant Medical History software 101–3, 104, 105
Institute of Medicine
 To Err Is Human 16
 Unequal Treatment report 54
insurance, algorithmic decision-making 259–60
internet
 open-source resources for LLM training 167
 patients' use of 93–5
 and telemedicine 67–9
 variable access to 67–8
 see also AI; computers; Google
Ioannidis, John 139
Iora Health 251–4, 255
Ireland, medical student survey 237–8
Isabel, diagnostic tool 197, 202–3

Jagger, Mick 156
Jha, Ashish 219–20
Jobs, Steve 165
jobs/employment 44, 156
 doctors' careers 20, 27, 30, 215
 effects of AI on 262, 266–7, 268–9
Johns Hopkins School of Medicine 92, 197–8
Jordan, Michael 123
Journal of the American Medical Association (*JAMA*) 16, 52
 on doctors' domination 90
 and doctors' use of AI 209
 on empathy 224

Journal of the Royal Society of Medicine 223
"Julie," and AIDS 230–1, 248

Kahneman, Daniel 163
 Thinking, Fast and Slow 160–2
Kaiser Permanente 63
Kalamazoo 42–4, 47
Kalanithi, Paul 235
 When Breath Becomes Air 220
Kasparov, Garry, and Kasparov's Law 195–6, 198, 210, 273
Kay, Adam 19, 234
 This Is Going to Hurt 18
Keeping Up with the Kardashians 16
Keillor, Garrison, Lake Wobegon 17
KevinMD blog 219–20
KFF health tracking poll 275
Khosla, Vinod, Sun Microsystems 28, 29, 32
"Koko" mental health chat service 249
Kostopoulou, Olga 201, 202–3
Kugelmass, Heather 146
Kuhn, Thomas, *The Structure of Scientific Revolutions* 20

The Lancet 53, 58, 82
languages, chatbots and 175
Lawson, Jennifer (Jen), case study 39–45, 47–53, 54, 56, 274
 and telemedicine 57, 59, 65, 73
Lawson, Tom (Professor Ernest Thomas), case study 77–80, 86, 97–8, 107, 114, 274
Leape, Lucian 90
Levy, Andrea 81, 83
Liaison Committee on Medical Education (US and Canada) 269
"Liam," case study 188–90, 199, 205, 274
licensing 25, 65–6
"LINC," early computer program 99–100, 109
LinkedIn 263
LLMs (large language model) chatbots 29, 93, 94, 106
 appeal of 208
 evaluation studies 209

human tutoring 167
reward-based tuning 167–8
self-supervision training 167
see also chatbots
lobbying 24–5
Logan Airport 57–8
Logg, Jennifer 199–200
Longwood Medical Area 215
Lucas, Gale 112
"luxury beliefs" 259–60, 276

McAfee, Andrew 28
McCauley, Robert 161
McDonald, Henry 11
McLuhan, Marshall 157
Madore, Kevin 225–6
Mahood, Sally 194
malpractice, legal suits 6
management styles 83–4
Manchester, University of 144
Marcus, Gary 32, 184
  Rebooting AI 29–30
The Markup 263
Marriott School of Business 112
Marsh, Henry 192, 194
  Admissions 193
  Do No Harm 18, 27, 218
Maryland, University of, School of Public Health 69
Massachusetts Institute of Technology (MIT) 100, 172
maternity services
  racial prejudice in UK 148–9
  US Black women 143, 146–7, 148–9
Maude, Jason 202–3, 207
Mayo Clinic, on video diagnosis 71–2
MD Anderson Cancer Center 28
ME (myalgic encephalomyelitis) 144–5
  AI analysis 173
  and discrimination 145
Med-PaLM-2, medical chatbot 209
Medicaid (US) 45, 147
medical data scientists 271
medical journals 140
medical profession
  culture of disrespect 91, 190–5
  ethical violations 26

  and hierarchy 191–2, 194
  and licensure 25, 65–6
  need for 270–73
  opacity 25–6
  powerplay problems 90–3, 191–2
  resistance to AI 266–7
  resistance to "second-level" professionals 25
  self-regulation 22–3
  see also doctors; nurses
medical reversals 139
medical students, and classism 146–7
medical training 21, 127–8
  about female biology 132–4
  and bad habits 128
  "board certification" (US) 129
  and changing treatments and practices 139–40
  and continuing education 140, 141–2, 170–1
  and doctors' socio-economic status 254–5
  and empathy 217, 223–4, 243–4
  failure to include AI 268–9
  "health coaches" 252–4
  "July effect" 127
  limitations of 131–9
  and practice 127–9
  variability of 128–9
  see also clinical research
medically unexplained symptoms 2, 151
Medicare (US) 26, 45, 61, 129
medicine, as dark art 158–62
MediKanren, AI tool 272–3
MedPage Today 182
Medscape 3, 217
Meehl, Paul, *Clinical versus Statistical Prediction* 200
Mello, Michelle 171
mental health
  access to care 146
  of doctors 3–4, 5, 226–8, 242
  and outdoor activity 117–18
  peer support 249, 256
  and telemedicine 60, 61, 72
Meta 24, 263
  "Blenderbot 3" chatbot 247

INDEX

Michigan, University of 184
Microsoft 267
Might, Bertrand 174
Might, Matt 173–4, 272–3
Miller, Glenn 43
mindLAMP app 117
Morrison, Van 156
multitasking 226–7, 242
muscular dystrophy 11
  myotonic dystrophy 137
Myrick, Keris 143–4, 241, 256

Nabla ambient AI 184
NASA, and telemedicine 58
Nash, David 70–1
National Academy of Sciences, Engineering, and Medicine 136
National Eating Disorder Helpline (US), "Tessa" chatbot 249
National Institute on Drug Abuse (US) 60
National Institutes of Health (NIH) 272
National Medical Association (US) 55
National Organization for Rare Disorders (US) 137
Nature Machine Intelligence 244
Nero, Emperor 145
Netflix 31
New England Journal of Medicine 169
  and changing practices 139–40
  and ChatGPT diagnoses 166
  on doctors' errors 18, 26
  on doctors' use of ChatGPT 208
  and GPT-4V testing 179
  on inconsistent treatment 124
  on internet access 68
  on telemedicine 59
  Warner Slack and 101
The New York Times 238, 243
Newman-Toker, David 2, 16
NGLY1 deficiency 174
NHS (National Health Service, UK) 20–1
  complaints about 97
  and complaints against doctors 96
  cyber-attack (2024) 262
  eConsult tool 105

racial prejudice 148
and telemedicine 60, 66, 72
"noise"
  aging 130
  AI and 169–70, 261
  clinical practice 124–5, 141
  medical expertise 167
  negative behaviors 128
  variable training 128–9, 170–1
Nuance, Ambient Experience app 267
nurse practitioners 25
nurses, relations with doctors 190–5

Oakland, California 149
Obermeyer, Ziad 176, 186–7
OECD 46, 114
Office of National Statistics (UK) 54
Ofri, Danielle 21, 168
"Oncotalk" course 243
O'Neil, Cathy, *Weapons of Math Destruction* 175
online dating 256–7
OpenAI
  and false data 182–4
  GPT-3.5 29
  GPT-4 29
  GPT-4o 29
  GPT-4V(ision) and difficult medical cases 178–9
  lobbying 24
  new models 32
  o1 series 186, 210, 211
  o3 series 186, 211
  Whisper 184
  see also chatbots; ChatGPT
opioid use, and telemedicine 60
Osler, William 216–17
outdoor activity 117–18
Oxford University 22, 234

pain
  AI assessment of 181–2
  gender differences 133
Palmer, Katie, *STAT* 263
paperwork 267
Parker, Dr. 78, 86
Parliamentary and Health Service Ombudsman 97

328

patients
 abuse of health staff 5
 access to AI 174
 access to healthcare 2, 5, 6–7
 access to records 23–4, 92–3
 and chatbots 35, 93–5, 113, 248–51, 274–5
 complaints against doctors 96
 embarrassment about symptoms 79–83, 107
 fear of physicians 95–7
 and feedback on treatment 142
 and health monitoring 114–16
 honesty with computers 107–8, 114
 interaction with computers 99–101, 105, 106–7, 109–10
 interactions with doctors 89–90, 154–5
 lies and deception 81, 107–8, 112–13
 passivity 150, 273
 studies of clinical empathy 220–1
 as subordinates 93–5, 147
 view of AI as complementary 196
 see also doctors
Patients Association (UK) 96
pattern recognition
 chatbots 171–3, 184
 doctors 172
paywalls 261
Peking University 231
Pemberton, Max, *Trust Me: I'm a (Junior) Doctor* 81
Perelman School of Medicine 64
Pew Research 55, 116, 148, 196, 256, 264
Pfizer 43
pharmaceutical companies 43
 and rare diseases 138–9
Phil, Dr. 98
philosophy 11, 110, 205, 228, 234
physician associates 25
Pierson, Emma 177, 180
Pignone, Michael 243–4
Pinel, Philippe 256
plagiarism, AI and 250
Playchess.com 195
Pope, Alexander 190

population, aging 4, 134–5
poverty 45–6, 129
Prasad, Vinay 139–40
prejudice, in healthcare 145–50
 against body weight 154–5
 against chronic conditions 150–2
 against disabled patients 152–3
 against sexual minorities 155–6
 against those with low incomes 49–50, 146–7, 149, 254
 see also bias; discrimination
presentation, self- 80
Presley, Lisa Marie 132
Prinz, Jesse, "Against empathy" 228, 234
privacy concerns 120, 207
 and data security 262–5
process managers, AI 271
Project Nightingale 263
Pronovost, Peter 16
pulse oximetry, and dark skin tones 136
Pussin, Jean-Baptiste 256

questionnaires, clinical 101–3, 104–5
Quinn, Dr., Medicine Woman 3

race, ethnicity, and segregation
 and access to healthcare 147–8
 AI identification of 175, 180
 chatbots and 176–7
 and clinical research 135–7
 and concordance of doctor and patient 149
 effect on empathy 231
 and facial recognition 175
 and genetic risks 180–1
 health disparities, US 54–6
 medical myths 136–7
 and stereotyping 160
 and telemedicine 64
radiologists versus AI 28, 182, 201
radiology, image readings 125
Rand Corporation 140
rare diseases
 AI identification of 173–4
 ChatGPT diagnosis of 171, 173

INDEX

definitions 137
diagnostic training for 137–9, 273
reciprocity, in relationships 150, 236, 256
Reddit 119, 167, 168
  AskDocs forum 247
Reis, Ben 120
Reiss, Helen 243
Revitalization Act (1993 US) 133
Riggare, Sara 115
robots 28
  anthropomorphized 111
Rodman, Adam 166, 186, 210, 274
Rodriguez, Jorge 64, 67, 68, 69, 70, 71
Rosenbaum, Edward, *A Taste of My Own Medicine* 14, 15
Roter, Debra 108–9
Royal College of General Practioners 96
Royal College of Physicians 66

Sackett, David 139
Schiff, Gordon 141
Schuetzler, Ryan 112–13
Schultz, Howard, Starbucks 123
Schwartz, William, "Medicine and the computer" (1970) 27
Scientific American 29
self-deception 15
Sellars, Wilfred 11
sexual harassment 194
Shem, Samuel, *The House of God* 91, 192
SHEPHERD, AI model 172–3
Shirky, Clay, "Shirky Principle" 266
sickle cell disease 135–6, 138
Silicon Valley 8, 10
Simmel, Marianne 111
Singapore, University of 94
Singh Ospina, Naykky 89–90
Slack, Warner 99, 101, 109
smart speakers (Siri, Alexa) 264
smartphones, for health monitoring 115–19
Smith, Richard 20
Snap 263
soap operas 251

social justice, and empathy 232–4
social media
  and health clues 119
  use of 115–16
Society to Improve Diagnosis in Medicine 16–17
Socrates 11
Solomon, Miriam 80
Southern California, University of 112, 249
Spock, Dr. 3, 222
Sproull, Lee 112
Stanford University 119, 139, 171, 176, 225, 238
Stephen, Zachary 195
stereotypes
  and biases in diagnosis 158–60
  "coalitional psychology" 160
  and racism 160
  social 159
stigmatizing language 92, 151, 154, 179
Street, Richard 150
suicide
  AI prediction of 119–20
  doctors and 4
Sun Microsystems 28
surveillance capitalism (Zuboff) 264–5
surveys
  of doctors' forecasts about job 203, 238, 268
  of generative AI use 34–5, 93, 113, 196, 264, 267
Susskind, Richard and Daniel 31–2
The Future of the Professions 22
Sweden
  access to records 92–3
  occupational heritability 255
  Work Environment Authority 93
Swift, Taylor 28
symptom denial 18–19
symptoms
  embarrassment about 79–83, 107
  gender differences 132–3
  and procrastination 82
Synnovis 262

System 1 and System 2 thinking 161–2, 172, 179, 185

TalkLife 244–5
Tan, Sharon 94
Tang, Liyan 183, 185
technology
　phases of 30–2
　possibilities of 27–8
　wearable 188–90
　see also AI; smartphones
Teledoc 60
telemedicine 57–61, 260
　AI-powered developers 271
　compared with in-person care 69, 70–2, 73
　and COVID-19 57–8, 60, 61, 66
　criticism of 58–9
　and data breaches 263–4
　diagnostic accuracy 71–2
　and disability 62–3
　doctors' resistance to 64–6
　and internet access 67–9
　limitations of 63
　and medical errors 70–1
　and mental health 60, 61, 72
　patients' view of 59–61
　by telephone 69, 71
　time and cost savings 63–4
　and vested interests 65–6
　by video 69, 71–2
"Tessa" chatbot 249
tethered cord syndrome (TCS) 172
Texas, University of, at Austin 183, 243
thinking
　fast (System 1) 161–2, 172, 179, 185
　and inbuilt skills 204
　slow (System 2) 161, 162, 185
Time magazine 30
Topol, Eric 30, 238
　Deep Medicine 217, 224
　Review, of medical education 269
Toronto, primary care study 146
Torous, John 116–18
transport
　distance and costs 48–50
　racial inequality (US) 56

triage 105, 152
　chatbots and 270–71
TrialGPT 173
Trivers, Robert 15
Truss, Liz 110
trust and transparency 113
　and data security 264–5
Tudor Hart, Julian 53
Turing Test 101
　ChatGPT and 106

United Nations 4
United States
　costs of healthcare 45–6
　diagnostic errors 1
　inconsistency of healthcare 124–5
　and internet access 67–8
　national security experts 200
　racial and ethnic health disparities 54–6
　transportation costs 48–9
　see also health insurance
Upjohn pharmaceutical company 43
Uppsala University Hospital 93
US Bureau of Labor Statistics 43, 46
US Center for Disease Control and Prevention 54
US Department of Veterans Affairs 61
UserTesting 93, 113

Verghese, Abraham 238
Vidal, Gore 88
Virginia, University of 17, 137
Viz magazine 246–7
Voltaire, *Candide* 15

Wachter, Robert 16
Wager, Anthony 225–6
waiting rooms 86–7
Walker, Jan 92
The Wall Street Journal 29, 263
Washington, University of 243, 244
The Washington Post 20, 95, 153
watches
　wearable technology 188–90
　see also smartphones
Waytz, Adam 227
WebMD 93

INDEX

weight discrimination 154–5
Weil, Kevin 9
Weizenbaum, Joseph 100–1, 120
Wenner, Allen 101–4, 105, 107–8, 185
Western Michigan University 44, 77
white coat hypertension 114
Williams, Serena 143
Wilson, Timothy, *Strangers to Ourselves* 17
Wisconsin, University of 99
women
  health disparities 132–3
  maternity services 143, 146–7, 148–9
  "medical gaslighting" 155
  and training about physiological differences 132–4
  treatment by male doctors 155
women doctors, performance 129–30, 155
Women and Equalities Committee (UK Parliament) 133
Wootton, David 268
  Bad Medicine 19
Workit Health 263
World Health Organization 4, 44
  on disability 53
  on empathy 217
  on telemedicine 70, 72

X/Twitter 119
Xenophanes 110
XPRIZE Foundation 28
Xu, Xiaojing 231

Yale School of Public Health 46

ZackS chess team 195–6, 198, 210, 273
Zappa, Frank 156
Zuboff, Shoshana 264
Zuckerberg, Mark 165, 247
Zulman, Donna 65, 67